Lunchuan Zhang

Hilbert C*- Modules and Quantum Markov Semigroups

Lunchuan Zhang
School of Mathematics
Renmin University of China
Beijing, Beijing, China

ISBN 978-981-99-8667-5 ISBN 978-981-99-8668-2 (eBook)
https://doi.org/10.1007/978-981-99-8668-2

Mathematics Subject Classification: 81R15, 60J46, 47D07, 47A53, 16D90

This Springer imprint is published by the registered company Springer Nature Singapore Pte Ltd.
The registered company address is: 152 Beach Road, #21-01/04 Gateway East, Singapore 189721, Singapore

Paper in this product is recyclable.

Preface

Nowadays the Hilbert C^*-module theory has become an important part of operator algebra. The concept of Hilbert C^*-module over commutative C^*-algebra was pioneered by I. Kaplansky in the 1950s, who used it as a tool to prove that derivations of type I AW^*-algebras are inner. The systematic study of Hilbert C^*-module over a general (non-commutative) C^*-algebra began with the work of W.L. Paschke and M.A. Rieffel in 1970s, respectively. In particular, Hilbert spaces and C^*-algebras are special Hilbert C^*-modules. Therefore Hilbert C^*-modules are the product of the combination of Hilbert spaces and C^*-algebras. Thus the basic theory of Hilbert C^*-modules is the result of the interaction between Hilbert spaces theory and C^*-algebras. Furthermore, with its applications in A.Connes' noncommutative geometry, S.L. Woronowicz's quantum group theory, Kasprov's KK-theory, Jones generalized index theory, and operator-valued free probability theory, the theory of Hilbert C^*-modules has developed vigorously. In recent years, it has been widely used in the study of quantum probability. These facts show that Hilbert C^*-modules theory has been developed in the process of applications.

This book is divided into three chapters. In Chap. 1 we shall introduce the basic theory of Hilbert C^*-modules, including the concepts of Hilbert C^*-modules and bounded module mappings, polar decomposition and Wold decomposition of bounded module mappings, bounded generalized inverse module mappings, module tensor product, and KSGNS construction; Chap. 2 describes mainly Kasprov's stability theory and Fredholm generalized index theory. It also includes the basic content of Morita equivalence and module framework. Chapter 3 is devoted to characterize a class of quantum Markov module operators semigroups and corresponding operator-valued Dirichlet forms. For this, we first introduce the concept of strong continuous (contractive) module operator semigroup, and obtain the Hille Yosida type theorem for characterizing the infinitesimal generators of module operator semigroup. Then the quantum Stone type theorem is given for describing strict continuous unitary module operator group. In addition, the abstract Cauchy problems based on Hilbert C^*-modules are discussed. Furthermore, we obtain the Beurling Denny criterion between a class of quantum Markov module operators semigroups and corresponding operator-valued Dirichlet forms. At the end of this chapter, we shall prove the equivalence between hypercontractivity for a class of quantum Markov semigroups and associated logarithmic Sobolev inequality based on probability gage space.

This book is self-contained as far as possible. Readers don't have much trouble reading this book as long as they have the basic knowledge of operator (matrix) algebras, and know a little about Markov semigroups and scalar-valued Dirichlet forms. As the writing feature, we pay attention to find ideas and clues for Hilbert C^*-modules theory from the Hilbert spaces and C^*-algebras, even from seemingly trivial examples. Meanwhile, the essential differences between Hilbert C^*-modules and Hilbert spaces are emphasized from different aspects. The first chapter is compiled by Associate Professor Xiaojing Wang, School of science, Beijing University of Civil Engineering and Architecture; the second and third chapters are written by Professor Lunchuan Zhang, School of mathematics, Renmin University of China.

I would like to express my deep gratitude to Professor E. Christopher Lance and Professor Maozheng Guo for their teachings. Thanks sincerely to Research Fellow Fuzhou Gong, Professor Qingwen Wang, and Professor Qingxiang Xu for their help. Thanks sincerely to Professor Zhiyong Zheng, Professor Yuanyuan Ke, Prof. Zhiyong Huang, Professor Dong Shen, and Professor Hao Jiang for their support. At this moment, I am particularly grateful to the editors for their concern and hard work for the publication of this book, and the valuable suggestions of the reviewers. Finally, I am pleased with the partial support of the National Natural Science Foundation of China (71771212, 12074428) for this work.

Beijing, China
December 2021

Lunchuan Zhang

Contents

Chapter 1
Basic Theory of Hilbert C^*-Modules

1.1 Hilbert C^*-Modules and Bounded Module Mappings

This chapter covers the basic concepts and methods of Hilbert C^*-modules. The main results of this chapter are multiplier theorem, polar decomposition and Wold decomposition, exterior and interior tensor product between Hilbert C^*-modules, and KSGNS representation theorem. In particular, the concept of bounded generalized inverse module mapping introduced in this chapter is used to characterize the factorization of bounded module mappings and solve module operator equations.

1.1.1 Hilbert C^*-Modules

Definition 1.1 Let A be a C^*-algebra, and let E be a complex linear space which is a right A-module, and $\lambda(xa) = (\lambda x)a = x(\lambda a)$ for $x \in E,\ a \in A, \lambda \in \mathbb{C}$. Then E is called pre-Hilbert A-module if there exists a mapping $< \cdot, \cdot >: E \times E \to A$ satisfying the following conditions:

(1) $< x,\ x >\geq 0$, and if $< x,\ x >= 0$ then $x = 0$;
(2) $< x,\ \lambda y >= \lambda < x,\ y >$ and $< x,\ y + z >=< x,\ y > + < x,\ z >$;
(3) $< x,\ ya >=< x,\ y > a$;
(4) $< x,\ y >=< y,\ x >^*$;

for all $x,\ y,\ z \in E$, $a \in A$ and λ in the complex field $\mathbb{C}$. In this case, the mapping $<, >$ is called a A-valued inner product. Hence a pre-Hilbert A-module is a A-valued inner product space.

Remark 1.1 The operator-valued and scalar-valued inner product spaces appearing in this book are conjugate linear in the first variable (i.e., the left variable), and linear in the second variable (i.e., the right variable), unless otherwise specified.

L. Zhang, *Hilbert C^*- Modules and Quantum Markov Semigroups*,
https://doi.org/10.1007/978-981-99-8668-2_1

Suppose that E is a pre-Hilbert A-module. Define a mapping $||\cdot||$ from E to $\mathbb{R}^+$ by the following way:

$$||\cdot|| : x \to ||x|| = || < x, \ x > ||^{1/2},$$

for all $x \in E$. In order to prove that $||\cdot||$ is a norm on E, we need the following Cauchy-Schwarz inequality:

Proposition 1.1 *Suppose that E is a pre-Hilbert A-module. Then*

$$< x, \ y >^* < x, \ y > \leq || < x, \ x > || < y, \ y >,$$

for all $x, \ y \in E$.

Proof Let f be any state on C^*-algebra A. Then it is easy to check that the following binary function is a semi-inner product on E:

$$E \times E \to \mathbb{C}, (x, \ y) \to f(< x, \ y >) \ (x, \ y \in E).$$

Since

$$< x, \ y >^* < x, \ y > = < y, \ x > < x, \ y > = < y, \ x < x, \ y >>,$$

it follows that

$$(f(< x, \ y >^* < x, \ y >))^2 \leq f(< y, \ y >) \cdot f(< x < x, \ y >, \ x < x, \ y >>),$$

from the Cauchy inequality in the case of scalar-value. Since

$$< x < x, \ y >, \ x < x, \ y >>$$

$$= < x, \ y >^* < x, \ x > < x, \ y >$$

$$\leq || < x, \ x > || < x, \ y >^* < x, \ y >,$$

from which it implies that

$$(f(< x, \ y >^* < x, \ y >))^2$$

$$\leq || < x, \ x > || f(< x, \ y >^* < x, \ y >) f(< y, \ y >).$$

Thus

$$\begin{aligned} f(< x, \ y >^* < x, \ y >) &\leq || < x, \ x > || f(< y, \ y >) \\ &= f(|| < x, \ x > || < y, \ y >), \end{aligned}$$

provided $f(< x, y >^* < x, y >) \neq 0$. In addition, the case of $f(< x, y >^* < x, y >) = 0$ is obviously true.

Finally, because the above f is an arbitrary state of C^*-algebra A. So

$$< x, y >^* < x, y > \leq || < x, x > || < y, y > .$$

Remark 1.2

(1) The following basic inequality in C^*-algebra is used in the above proof: If a and b are self-adjoint elements in a C^*-algebra A and $a \leq b$, then $c^*ac \leq c^*bc$ for all c in A.
(2) Proposition 1.1 can also be proved by using the definition of pre-Hilbert A-module and the basic properties of positive cone of C^*-algebra. For details, one can refer to E.C. Lance [13] Proposition 1.1.

From Proposition 1.1, we immediately get the following corollary:

Corollary 1.1 *Suppose that E is a pre-Hilbert A-module. Then*

$$|| < x, y > ||^2 \leq || < x, x > || \cdot || < y, y > ||,$$

for all $x, y \in E$

Set $||x|| = || < x, x > ||^{1/2}$ $(x \in E)$. By Corollary 1.1, it is easy to see that $|| \cdot ||$ is a norm on E. Moreover, we have

$$||x|| = sup\{|| < x, y > || : y \in E, ||y|| \leq 1\}.$$

Definition 1.2 A pre-Hilbert A-module E over a C^*-algebra A which is complete with respect to the above norm is called a Hilbert C^*-module over C^*-algebra A, in this case, E as linear space is a Banach space. In short, it is called a Hilbert A-module, or Hilbert C^*-module, when no confusion occurs.

A closed subspace F of a Hilbert A-module E becomes a Hilbert C^*-closed submodule of E if it is a right A module itself, namely, $F \cdot A \subseteq F$, it is usually called a closed submodule of E, in short. Thus $\{0\}$ and E are closed submodules of E, which are called trivial closed submodules.

Example 1.1 Let A be a C^*-algebra. Define a map given by:

$$< \cdot, \cdot >: A \times A \to A, (a, b) \to < a, b >= a^*b,$$

for all $a, b \in A$. It is easily checked that the above mapping satisfies the conditions in Definition 1.2. Therefore A itself becomes a Hilbert C^*-module whose norm as Hilbert module is identified with the uniform operator norm on A as C^*-algebra.

From the above observation, we see that every closed right ideal B of A is a closed submodule of A, where the A-value inner product on B is given by the following way

$$< a,\ b >= a^* b \in B \subset A,$$

for all $a,\ b \in B$.

Remark 1.3

(1) If C^*-algebra A reduces to the complex field $\mathbb{C}$, then the Hilbert module E in the Definition 1.2 becomes a Hilbert space. Thus, Hilbert space can be regarded as a special case of Hilbert C^*-module. Hence, the theory of Hilbert space can enlighten the establishment and development of Hilbert C^*-module theory. But we shall see later that there are many essential differences between the general Hilbert C^*-modules and Hilbert spaces.

(2) Similar to the complex Hilbert space case, the operator valued inner product of Hilbert C^*-module also satisfies polarization identity: Suppose that E is a Hilbert A-module. Then the following identity holds:

$$< x,\ y >= \frac{1}{4}\{< x + y,\ x + y > - < x - y,\ x - y > +$$

$$+ i < ix + y,\ ix + y > - < ix - y,\ ix - y >\},$$

for all $x,\ y \in E$.

(3) Similarly, we can also define the concept of left module, and then we shall introduce the concept of bimodule.

Definition 1.3 Let E be a Hilbert A-module and subset $\{x_i\}_{i\in I} \subseteq E$, where I is the index set. If the set of finite A-linear combinations of elements in $\{x_i\}_{i\in I}$ which denoted by $span\{x_i\}_{i\in I}$ is dense in E, then it is called the generating set of E. Thus, E is said to be countably generated if the index set I is countable, and if I is finite then E is (topologically) finitely generated. In particular, if there are finite elements $x_1,\ x_2,\ \cdots,\ x_n$ in E such that $E = span\{x_1,\ x_2,\ \cdots,\ x_n\}$, then E is called algebraically finitely generated Hilbert C^*-module.

Example 1.2 Let A be a C^*-algebra, and let $l^2(A) = \{(a_i)_{i=1}^{\infty} \in \prod_1^{\infty} A : \sum_{i=1}^{n} a_i^* a_i$ is norm convergent in A, $n \in \mathbb{N}\}$. Define a A-valued inner product as follows:

$$< \cdot,\ \cdot >:\ l^2(A) \times l^2(A) \to A,\ < (a_i)_{i=1}^{\infty},\ (b_i)_{i=1}^{\infty} >= \sum_{i=1}^{\infty} a_i^* b_i.$$

Thus $l^2(A)$ is a Hilbert C^*-module.

Remark 1.4 Let $(a_i)_{i=1}^{\infty}$ be in $\prod_1^{\infty} A$. Since $||\sum_{i=1}^{n} a_i^* a_i|| \leq \sum_{i=1}^{n} ||a_i||^2$, $\sum_{i=1}^{\infty} a_i^* a_i$ is norm convergent in A when $\sum_{i=1}^{\infty} ||a_i||^2$ converges. But the opposite is not true in generally.

Example 1.3 Let $A = C_0(1, +\infty)$ be the set of all continuous functions on $(1, +\infty)$ vanishing at infinity, then it is an abelian C^*-algebra with norm $||x|| = \sup\{|x(t)| : t \in (1, +\infty)\}$. Choose a sequence of functions in A as follows:

$$x_n(t) = \frac{\sqrt{n}}{t^{\frac{n}{2}}}, t \in (1, +\infty).$$

By the power series theory in mathematical analysis, we have

$$\sum_{n=1}^{\infty} x_n^*(t) x_n(t) = \sum_{n=1}^{\infty} \frac{\sqrt{n}}{t^{\frac{n}{2}}} \cdot \frac{\sqrt{n}}{t^{\frac{n}{2}}} = \sum_{n=1}^{\infty} \frac{n}{t^n},$$

which is pointwise convergence on $(1, +\infty)$. Since

$$||x_n|| = \sup_{t \in (1, +\infty)} |x_n(t)| = \sup_{t \in (1, +\infty)} \frac{\sqrt{n}}{t^{\frac{n}{2}}} = \sqrt{n},$$

which implies that $\sum_{n=1}^{\infty} ||x_n||^2 = \sum_{n=1}^{\infty} n$ is divergence.

Remark 1.5 Set $E_A = \{(a_i)_{i=1}^{\infty} \in \prod_1^{\infty} A : \sum_{i=1}^{\infty} ||a_i||^2 < \infty\}$. Then $l^2(A) = E_A$ if and only if A is a finite dimensional C^*-algebra. Refer to [6] and [9] for details.

It is well known that orthogonality is the feature of inner product spaces. Analogously, the concept of orthogonality can be transplanted into Hilbert A-modules as follows:

Definition 1.4 Let x and y be elements in a Hilbert A-module E. If $< x, y >= 0$ then they are called orthogonal, or perpendicular, which is denoted by $x \perp y$. Furthermore, given a closed submodule F of E and x in E, if $< x, y >= 0$ for every y in F, then we say that x is perpendicular to F, which is denoted by $x \perp F$. Moreover, F is said to be orthogonal complemented whenever $F \oplus F^{\perp} = E$, where $F^{\perp} = \{x \in E : x \perp F\}$.

Remark 1.6

(1) In this book, we only use the concept of orthogonal complemented module, so we abbreviate frequently orthogonal complemented module to complemented module;
(2) Every closed subspace of Hilbert space can be complemented, but Hilbert C^*-modules generally have no such property, which is the essential difference between them.

Example 1.4 Let H be an infinite dimensional separable Hilbert space and $B(H)$ the C^*-algebra consisting of all bounded linear operators on H, $K(H)$ the compact operator ideal in $B(H)$. Since $K(H)$ is dense in $B(H)$ with respect to the weak operator topology, $K(H)^{\perp} = \{0\}$. It follows that

$$K(H) \oplus K(H)^{\perp} = K(H) \neq B(H),$$

which shows that $K(H)$ is not complemented.

As we shall see later, complemented closed submodules play a key role in the factorization of bounded module maps.

Definition 1.5 Let A be a unital C^*-algebra, and let $A^n = \underbrace{A \oplus A \oplus \cdots \oplus A}_{n}$ be the free A-module. A Hilbert A-module E is called a finitely generated projective module, which means that if there is a natural number n such that E is an orthogonal complement submodule of A^n.

Remark 1.7 Countably generated Hilbert C^*-modules, algebraic finitely generated modules, and finitely generated projective modules are all important roles in the generalized Fredholm index theory in the second chapter.

At the end of this section, we give a basic and important property of Hilbert C^*-modules:

Proposition 1.2 *Suppose that E is a Hilbert A-module. Then $\overline{< E, E >}$ (closure in the norm topology in A) is a two-sided ideal in A, and $E < E, E >$ is dense in E with respect to the norm on E, where $< E, E >= span\{< x, y >: x, y \in E\}$.*

Proof From the meaning of Hilbert C^*-module, it is clear that

$$a < x, y >=< xa^*, y >\in< E, E >,$$

and

$$< x, y > a =< x, ya >\in< E, E >,$$

for all $x, y \in E$ and $a \in A$. It follows that $\overline{< E, E >}$ is a two-sided ideal in A. If $\{u_i\}$ is an approximate unit of $\overline{< E, E >}$, then we have

$$||x - xu_i||^2 = || < x - xu_i, x - xu_i > ||$$

$$= || < x, x > -u_i < x, x > - < x, x > u_i + u_i < x, x > u_i||$$

$$\leq || < x, x > - < x, x > u_i|| + ||u_i < x, x > u_i - u_i < x, x > ||$$

$$\leq || < x, x > - < x, x > u_i|| + || < x, x > u_i - < x, x > || \longrightarrow 0.$$

Hence $E < E, E >$ is dense in E.

Remark 1.8

(1) If $\overline{< E, \ E >} = A$, then E is called a full Hilbert C^*-module;
(2) From Proposition 1.2, we see that EA is dense in E. So if A has unit 1 then $x1 = x$ for every $x \in E$; If A has no unit, let $\widetilde{A} = A \oplus \mathbb{C}$, then put $x1 = x$ for every x in E, hence E becomes a Hilbert C^*-module over C^*-algebra $\widetilde{A}$, where $1 = (0, 1)$ is the unit of $\widetilde{A} = A \oplus \mathbb{C}$.

1.1.2 Bounded Module Mappings

Suppose that E and F are both Hilbert C^*-modules over a C^*-algebra A. Then a linear mapping $T : E \to F$ is called A-linear mapping from E to F if it satisfies the following condition:

$$T(x \ a) = T(x) \ a \ (x \in E, a \in A).$$

Because E and F are still Banach spaces, we can define the operator norm on T as follows:

$$||T|| = \sup\{||Tx|| : ||x|| \leq 1\}.$$

If $||T|| < \infty$, then T is called bounded module mapping. The set of all bounded module mappings from E to F is denoted by $B(E, \ F)$.

Definition 1.6 Let E and F be Hilbert C^*-A-modules. Then a mapping $T : E \to F$ is said to be adjointable if there exists a mapping $T^* : F \to E$ satisfying

$$< Tx, \ y >=< x, \ T^* y >,$$

for all $x \in E$ and $y \in F$. Such a mapping T^* is then called the adjoint of T.

We shall prove that the adjointable mapping T must be bounded A-linear mapping. Indeed, since

$$< T(x \ a), \ y >=< x \ a, \ T^*(y) >$$

$$= a^* < x, \ T^*(y) >= a^* < T(x), \ y >$$

$$=< T(x) \ a, y > \ (x \in E, \ y \in F, \ a \in A).$$

So that $T(x \ a) = T(x) \ a$, which shows that T is a A-linear mapping.

Now we show that T is bounded. For this, define a linear mapping given by

$$T_x : F \to A, \ y \to < Tx, \ y > \ (y \in F),$$

for each fixed $x \in E$. Since

$$||T_x(y)|| = || < Tx, \ y > ||$$

$$=< x, \ T^*y > \leq ||x|| \cdot ||T^*y||,$$

which means that T_x is pointwise bounded for all y in F. Therefore T_x is uniform bounded by the Banach-Steinhaus Theorem (viz., the uniform boundedness principle). Hence there is a positive constant M that independent of y such that

$$||T_x|| \leq M||x||.$$

From this, we can get

$$||T_x(y)|| = || < TX, \ y > || \leq M||x|| \cdot ||y||.$$

Then put $y = Tx$, we have

$$||TX||^2 = || < Tx, \ Tx > || \leq M||x|| \cdot ||Tx||.$$

Hence

$$||Tx|| \leq M||x||.$$

So that T is bounded.

By $L(E, \ F)$, we denote the set of all adjointable mappings from E to F. From the above proof, it is known that $L(E, \ F) \subseteq B(E, \ F)$. When E and F are degenerated into Hilbert spaces, we have $L(E, \ F) = B(E, \ F)$, since every bounded linear operator between Hilbert spaces has adjoint operator. However, $L(E, \ F) \neq B(E, \ F)$ for Hilbert C^*-modules, in general. This is also the essential difference between Hilbert C^*-modules and Hilbert spaces.

We now consider an infinite dimensional separable Hilbert space H. Let $E = K(H), \ F = B(H)$, then E and F can be regarded as Hilbert C^*-modules over the C^*-algebra $B(H)$. Let $j : E \to F$, be the embedding mapping. Then j is not adjointable. Otherwise, if $t : F \to E$, is its adjoint mapping of j, then

$$< jx, \ y >=< x, \ ty > \ (x \in E, \ y \in F).$$

Since $jx = x$ for all $x \in E$, it follows that

$$< x, \ y >=< jx, \ y >=< x, \ ty > .$$

Hence

$$< x, \ ty - y >= 0.$$

Therefore

$$t\,y - y \in E^{\perp} = K(H)^{\perp} = \{0\}.$$

Above proof shows that $ty = y$ for all $y \in F = B(H)$, that is, t is the identity mapping. So that $K(H) = B(H)$, this contradicts to that $K(H)$ is a proper subset of $B(H)$.

Remark 1.9

(1) In particular, if $E = F$, then we write $B(E)$ and $L(E)$ instead of $B(E,\ F)$ and $L(E,\ F)$, respectively.
(2) It is easy to prove that $B(E,\ F)$ and $L(E,\ F)$ are Banach spaces with respect to the operator norm $||T|| = \{||Tx|| : ||x|| \leq 1\}$. Moreover, $B(E)$ is a Banach algebra and $L(E)$ is a C^*-algebra.

We next prove that $L(E)$ is a C^*-algebra. It is clear that $L(E)$ is a Banach space with respect to the operator norm, and for any $T_1,\ T_2 \in L(E)$ then T_1T_2 has adjoint $T_2^*T_1^*$. Thus $T_1T_2 \in L(E)$. Define the involution $*$ (or $*$-operation, in short) given by

$$* : L(E) \to L(E),\ T \to T^*\ (T \in L(E)).$$

Since

$$||T_1T_2(x)|| = ||T_1(T_2(x))|| \leq ||T_1|| \cdot ||T_2x|| \leq ||T_1||||T_2||||x||,$$

it follows that

$$||T_1T_2|| \leq ||T_1||||T_2||.$$

Also

$$|| < Tx,\ y > || = || < x,\ T^*y > || \leq ||x|| \cdot ||T^*y||$$

$$\leq ||T^*||||x||||y||.$$

Therefore

$$||Tx|| = sup\{|| < Tx,\ y > || : ||y|| \leq 1\} \leq ||T^*||||x||,$$

which implies that $||T|| \leq ||T^*||$. Because $(T^*)^* = T$ by symmetry, so we have also $||T^*|| \leq ||T||$. Hence $||T|| = ||T^*||$, which shows the $*$-operation is an isometric mapping. Therefore $L(E)$ is a Banach $*$ algebra.

Finally, we shall prove the C^*-identity: $||T^*T|| = ||T||^2$ $(T \in L(E))$. From the above proof, we see that

$$||T^*T|| \leq ||T^*|| \cdot ||T|| = ||T||^2.$$

Therefore, in order to prove the above C^*-identity, we need only prove the inverse inequality:

$$||T^*T|| \geq ||T||^2.$$

Indeed, since

$$||Tx||^2 = || < Tx,\ Tx > || = || < x,\ T^*Tx > ||$$

$$\leq ||T^*Tx|| \cdot ||x|| \leq ||T^*T|| \cdot ||x||^2.$$

from which it implies that

$$||T|| = \sup\{||Tx|| : ||x|| \leq 1\}$$

$$\leq \sup\{||T^*T||^{1/2}||x|| : ||x|| \leq 1\}$$

$$= ||T^*T||^{1/2}.$$

This proves our claim.

Proposition 1.3 *Suppose that E and F are Hilbert C^*-modules over a C^*-algebra A, and suppose that T is in $L(E,\ F)$. Then we have*

$$< Tx,\ Tx > \leq ||T||^2 < x,\ x >,$$

for all x in E

Proof Assume that f is any given state on the C^*-algebra A. Then E becomes a semi-inner product space when endowed with the following semi-inner product

$$(x,\ y) \to f(< x,\ y >)\ (x,\ y \in E).$$

Using Cauchy inequality repeatedly, we have

$$f(< T^*Tx,\ x >) \leq [f(< T^*Tx,\ T^*Tx >)]^{1/2} \cdot [f(< x,\ x >)]^{1/2}$$

$$= [f((T^*T)^2x,\ x)]^{1/2} \cdot [f(< x,\ x >)]^{1/2}$$

$$\leq [f(< (T^*T)^2x,\ (T^*T)^2x >)]^{1/4} \cdot [f(< x,\ x >)]^{\frac{1}{2}+\frac{1}{4}}$$

$$\cdots$$

$$\leq [f(< (T^*T)^{2^n}x,\ x >)]^{2^{-n}} \cdot [f(< x,\ x >)]^{\frac{1}{2}+\frac{1}{4}+\cdots+\frac{1}{2^n}}$$

$$\leq [||T^*T||^{2^n} f(< x,\ x >)]^{2^{-n}} \cdot [f(< x,\ x >)]^{1-2^{-n}}$$

$$\leq ||T^*T|| \cdot (||x||^2)^{2^{-n}} \cdot [f(< x,\ x >)]^{1-2^{-n}}.$$

It follows that

$$f(< T^*Tx,\ x >) \leq ||T^*T|| f(< x,\ x >) = f(||T||^2 < x,\ x >),$$

when $n \to \infty$. Therefore

$$< T^*Tx,\ x > \leq ||T||^2 < x,\ x >,$$

since f is an arbitrary state on A. Hence $< Tx,\ Tx > \leq ||T||^2 < x,\ x >$, as desired.

In Hilbert spaces theory and C^*-algebra, compact operator ideal plays an important role, whose is generated by rank one operators. In what follows, the rank one operators will be generalized to Hilbert C^*-modules:

Let E and F be Hilbert C^*-modules over a C^*-algebra A. For any fixed $x \in E$ and $y \in F$, define

$$\theta_{x,\ y} : F \to E,\ \theta_{x,\ y}(z) = x < y,\ z > \quad (z \in F).$$

We have then

$$< \theta_{x,\ y}(z),\ w > = < x < y,\ z >,\ w >$$

$$= < y,\ z >^* < x,\ w > = < z,\ y >< x,\ w >$$

$$= < z,\ y < x,\ w >> = < z,\ \theta_{y,\ x}(w) > \quad (w \in E),$$

this shows that $(\theta_{x,\ y})^* = \theta_{y,\ x}$. So $\theta_{x,\ y} \in L(F,\ E)$.

We denote by $K(F,\ E)$ the closed linear subspace of $L(F,\ E)$ spanned by $\{\theta_{x,\ y} : x \in E,\ y \in F\}$, and write $K(E)$ for $K(E,\ E)$. By the meaning of $\theta_{x,\ y}$ it is easy to check that the following properties hold:

Suppose that E, F and G are Hilbert C^*-modules over a C^*-algebra A. Then we have

$$T\theta_{x,\ y} = \theta_{Tx,\ y},$$

and

$$\theta_{x,\ y}S = \theta_{x,\ S^*y},$$

for all $T \in L(E,\ G)$ and $S \in L(G,\ F)$. In particular,

$$\theta_{x,\ y}\theta_{u,\ v} = \theta_{x<y,\ u>,v} = \theta_{x,\ v<u,\ y>}\ (x \in E,\ \ y \in F,\ \ u \in F,\ \ v \in G).$$

Thus $K(E)$ is a closed two-sided ideal of $L(E)$. Therefore there is an inclusive relationship among $K(E)$, $L(E)$ and $B(E)$ as follows:

$$K(E) \subseteq L(E) \subseteq B(E).$$

Remark 1.10 When Hilbert C^*-module E degenerates to Hilbert space, $\theta_{x,\ y}$ becomes a rank one operator on Hilbert space, and $K(E)$ is the compact operator ideal. Therefore, $K(E)$ is an analogue of compact operator ideal. However, the elements in $K(E)$ are not compact operators, and $K(E)$ is not unique as a closed two-sided ideal of $L(E)$ in general. Nevertheless, $K(E)$ plays an important role in the study of Hilbert C^*-modules.

Example 1.5 Let A be a C^*-algebra which is regarded itself as a Hilbert C^*-module. Then $K(A)$ and A are $*$-isomorphic, namely, $K(A) \cong A$.

Indeed, we can define a linear mapping $\Phi : span\{\theta_{a,b} : a, b \in A\} \to A$ by the following way

$$\theta_{a,\ b} \to ab^*\ (a,\ b \in A).$$

Since

$$||\sum_{i=1}^{n} \theta_{a_i,\ b_i}(c)|| = ||\sum_{i=1}^{n} a_i b_i^* c|| = ||(\sum_{i=1}^{n} a_i b_i^*)c||,$$

for all $a_1,\ a_2,\ \cdots,\ a_n,\ b_1,\ b_2,\ \cdots,\ b_n$ and c in A. It follows that

$$||\sum_{i=1}^{n} \theta_{a_i,\ b_i}|| = ||\sum_{i=1}^{n} a_i b_i^*||,$$

which shows that the above mapping Φ is isometric. Note that $\overline{span\{\theta_{a,\,b} : a,\; b \in A\}} = K(A)$. Thus Φ can be extended by continuity to an isometry from $K(A)$ to A, which is still denoted as Φ.

We prove next that the $*$-operation is preserved. Since

$$\Phi(\theta_{a,\,b}\theta_{c,\,d}) = \Phi(\theta_{a<b,\,c>,\,d})$$

$$= a < b,\; c > d^* = ab^*cd^*$$

$$= (ab^*)(cd^*) = \Phi(\theta_{a,\,b})\Phi(\theta_{c,\,d})\;(a, b, c, d \in A).$$

Also

$$\Phi(\theta_{a,\,b}^*) = \Phi(\theta_{b,\,a})$$

$$= ba^* = (ab^*)^* = (\Phi(\theta_{a,\,b}))^*,$$

Hence Φ is $*$ preserving isomorphism.

Finally, to show that Φ is onto, let $\{u_\lambda\} \subset A$ be an approximate unit of A, we have then

$$||\Phi(\theta_{a,\,u_\lambda}) - a|| = ||au_\lambda - a|| \to 0\; for\; all\; a\; in\; A,$$

which shows that $\Phi(\theta_{a,u_\lambda})$ is Cauchy net in A. Recalling that Φ is isometric from the above proof. Thus $\{\theta_{a,\,u_\lambda}\}$ is a Cauchy net in $K(A)$. Therefore $\theta_{a,\,u_\lambda} \to k_a \in K(A)$. It implies that

$$||a - \Phi(k_a)|| = ||\Phi(\theta_{a,\,u_\lambda}) - \Phi(k_a) + a - \Phi(\theta_{a,\,u_\lambda})||$$

$$\leq ||\Phi(\theta_{a,\,u_\lambda}) - \Phi(k_a)|| + ||a - \Phi(\theta_{a,\,u_\lambda})||$$

$$= ||\theta_{a,\,u_\lambda} - k_a|| + ||\Phi(\theta_{a,\,u_\lambda}) - a|| \to 0,$$

which means that $\Phi(k_a) = a$, so that Φ is onto as desired.

Remark 1.11

(1) If A has unit, then it is easy to see that

$$A \cong K(A) = L(A) = B(A).$$

(2) If A is a C^*-algebra and n is a positive integer, we write $M_n(A)$ for the C^*-algebra of $n \times n$ matrices over A. Suppose that $E_1, E_2, \cdots, E_n$ are Hilbert A-modules, then their direct $\oplus_{i=1}^n E_i$ is also a Hilbert A-module when endowed with the following A-valued inner product:

$$< x, y > = \sum_{i=1}^{n} < x_i, y_i > \quad (x = (x_1, x_2, \cdots, x_n),$$

$$y = (y_1, y_2, \cdots, y_n)).$$

In the later we rewrite $\oplus_{i=1}^n E_i$ as E^n. It is easily checked that $L(E^n)$ is naturally identified with $M_n(L(E))$ and $K(E^n)$ with $M_n(K(E))$, respectively.

Proposition 1.4 *Suppose that E is a Hilbert A-module. Then*

$$||\theta_{\xi,\,\xi}|| = || < \xi, \xi > || \textit{ for all } \xi \in E.$$

Proof When ξ is in E. Then we have

$$\theta_{\xi,\,\xi} = sup\{||\theta_{\xi,\,\xi}(\eta)|| : ||\eta|| \le 1\}$$

$$= \sup\{||\xi < \xi, \eta > || : ||\eta|| \le 1\}$$

$$\le ||\xi|| \cdot \sup\{|| < \xi, \eta > || : ||\eta|| \le 1\}$$

$$\le ||\xi|| \cdot ||\xi|| = ||\xi||^2 = || < \xi, \xi > ||;$$

In what follows, let $\eta = \frac{\xi}{||\xi||}$ if $\xi \neq 0$, then

$$||\theta_{\xi,\,\xi}|| \ge ||\theta_{\xi,\,\xi}(\frac{\xi}{||\xi||})||$$

$$= \frac{1}{||\xi||}||\xi < \xi, \xi > ||$$

$$= \frac{1}{||\xi||}(||\xi < \xi, \xi > ||^2)^{1/2}$$

$$= \frac{1}{||\xi||}(|| < \xi, \xi >^3 ||^{1/2}).$$

Since $< \xi, \xi >> 0$, then it follows from the function calculus of the positive elements of C^*-algebra that

$$|| < \xi, \xi >^3 || = || < \xi, \xi > ||^3.$$

Combining this with the earlier inequality, one can obtain

$$||\theta_{\xi,\ \xi}|| \geq \frac{1}{||\xi||}|| < \xi,\ \xi > ||^{3/2}$$

$$= \frac{1}{||\xi||}\left(||\xi||^2\right)^{3/2} = ||\xi||^2,$$

$$= || < \xi,\ \xi > ||.$$

The above inequality is obviously valid for $\xi = 0$. Therefore

$$||\theta_{\xi,\ \xi}|| = || < \xi,\ \xi > ||.$$

Corollary 1.2 *The conditions are the same as Proposition* 1.4. *Then for any* $x_1,\ x_2,\ \cdots,\ x_n,\ y_1,\ y_2,\ \cdots,\ y_n \in E$, *we have*

$$||\sum_{i=1}^{n}\theta_{x_i,\ y_i}|| = ||\left(< x_i,\ x_j >\right)_{ij}^{1/2}\left(< y_i,\ y_j >\right)_{ij}^{1/2}||,$$

where the matrices $(< x_i,\ x_j >)_{ij}$ *and* $(< y_i,\ y_j >)_{ij}$ *are in matrix algebra* $M_n(A)$. *In particular, if* $E = A$ *then*

$$||\sum_{i=1}^{n}x_i y_i^*|| = ||\left(x_i^* x_j\right)_{ij}^{1/2}\left(y_i^* y_j\right)_{ij}^{1/2}||.$$

Proof Let $E^n = E \oplus E \oplus \cdots \oplus E$. Then E^n becomes a Hilbert C^*-module over $M_n(A)$ equipped with the following $M_n(A)$-valued inner product:

$$< x,\ y >= (< x_i,\ y_j >)_{ij} \in M_n(A),$$

for all $x = (x_1,\ x_2,\ \cdots,\ x_n)$ and $y = (y_1,\ y_2,\ \cdots,\ y_n)$ in E^n. Therefore

$$\theta_{x,\ y}(z) = x < y,\ z >= \left(\sum_{i=1}^{n}\theta_{x_i,\ y_i} \otimes I\right)(z),$$

which shows that $\theta_{x,\ y} = \sum_{i=1}^{n} \theta_{x_i,\ y_i} \otimes I(z = (z_1,\ z_2,\ \cdots,\ z_n) \in E^n,)$, where I stands for the unit element of $M_n(A)$, that is, the identity matrix. Thus, it results from Proposition 1.4 that

$$||\sum_{i=1}^{n} \theta_{x_i,\ y_i}|| = ||\theta_{x,\ y}||$$

$$= || < x,\ x >^{1/2} < x,\ x >^{1/2} ||$$

$$= || \left(< x_i,\ x_j >\right)_{ij}^{1/2} \left(< y_i,\ y_j >\right)_{ij}^{1/2} ||.$$

Riesz Frechet representation theorem is the cornerstone of Hilbert space theory, which shows that each bounded linear functional f on Hilbert space can be written in the form of inner product. Concretely, let f be a bounded linear functional on a Hilbert space H, then there exists a unique y_f in H such that $f(x) =< y_f,\ x >$ for all $x \in H$, and $||f|| = ||y_f||$. The essence of this result is the selfduality of Hilbert Space, namely, $H \cong H^*$, where H^* is the dual space of H, that is the set of all bounded linear functionals on H. However, bounded module maps generally do not have this property, which is the essential difference between Hilbert spaces and Hilbert C^*-modules.

For example, we consider also an infinite separable Hilbert space H and its compact operator ideal $K(H)$, then $K(H)$ becomes a Hilbert C^*-module over C^*-algebra $B(H)$ with the $B(H)$-valued inner product by

$$< a,\ b >= a^*b,$$

for all $a,\ b \in K(H)$. It is easy to check that the identity mapping $i : K(H) \to K(H) \subset B(H),\ x \to x\ (x \in K(H))$, cannot be written in the form of $B(H)$-valued inner product. The details are left to the reader.

In the framework of Hilbert C^*-modules, the classical Riesz-Frechet representation theorem can be generalized from two aspects:

On the one hand, $K(E,\ A)$ can be regarded as the analogue of dual space of Hilbert space, because the module mapping in $K(E,\ A)$ can be written in the form of A-valued inner product:

Proposition 1.5 *Suppose that E is a Hilbert A-module and T is in $K(E,\ A)$. Then there exists a unique $y_T \in E$ such that $T(x) =< y_T,\ x >$ for all $x \in E$, and $||T|| = ||y_T||$.*

Proof We first prove the following claim:

There exists an isometric isomorphism from $K(E,\ A)$ onto E as Banach spaces.

For this, construct a linear mapping $\Phi : K(E,\ A) \to E$ given by

$$\Phi : \sum_{i=1}^{n} \theta_{a_i,\ x_i} \to \sum_{i=1}^{n} x_i a_i^*,$$

for all $x_1, x_2, \cdots, x_n$ in E and $a_1, a_2, \cdots, a_n$ in A. Since

$$\sum_{i=1}^{n} \theta_{a_i, x_i}(y) = \sum_{i=1}^{n} a_i < x_i, y >$$

$$= \sum_{i=1}^{n} < x_i a_i^*, y > = < \sum_{i=1}^{n} x_i a_i^*, y > \quad for\ all\ y\ in\ E,$$

from which it implies that

$$||\sum_{i=1}^{n} \theta_{a_i, x_i}|| = ||\sum_{i=1}^{n} x_i a_i^*||.$$

Thus, Φ is isometric from $span\{\theta_{a,x} : a \in A, x \in E\}$ to E. Note that

$$\overline{span\{\theta_{a, x} : a \in A, x \in E\}} = K(E, A),$$

Therefore Φ can be extended to an isometric mapping from $K(E, A)$ to E, which is still denoted as Φ.

We prove next that Φ is surjective. Let $\{u_\lambda\}$ be an approximate unit of A. Since for any $x \in E$ we have

$$||xu_\lambda - x|| = || < xu_\lambda - x, xu_\lambda - x > ||^{1/2}$$

$$= ||u_\lambda < x, x > u_\lambda - u_\lambda < x, x > - < x, x > u_\lambda + < x, x > ||^{1/2} \longrightarrow 0,$$

which shows that $\{xu_\lambda\}$ is a Cauchy net in E. It follows that $\{\theta_{u_\lambda, x}\}$ is a Cauchy net in $K(E, A)$ since Φ is isometric whose converges to $T_x \in K(E, A)$. Also

$$||\Phi(T_x) - x|| = ||\Phi(T_x) - \Phi(\theta_{u_\lambda, x}) + \Phi(\theta_{u_\lambda, x}) - x||$$

$$\leq ||\theta_{u_\lambda, x} - T_x|| + ||xu_\lambda - x|| \to 0.$$

Thus $\Phi(T_x) = x$, so Φ is onto.

Now we can prove the conclusion. Since $\overline{span\{\theta_{a, x} : a \in A, x \in E\}} = K(E, A)$. For any given $T \in K(E, A)$, there exist a Cauchy net $\{T_\lambda\}_{\lambda \in \Gamma}$ (where Γ is the index set) $\subset span\{\theta_{a, x} : a \in A, x \in E\}$ such that $||T_\lambda - T|| \to 0$. Concretely, there exist $a_1^\lambda, a_2^\lambda, \cdots, a_{n(\lambda)}^\lambda$ in A, and $x_1^\lambda, x_2^\lambda, \cdots, x_{n(\lambda)}^\lambda$ in E, where $n(\lambda)$ is a natural number that depending with λ, such that

$$T_\lambda = \sum_{i=1}^{n(\lambda)} \theta_{a_i^{(\lambda)}, x_i^{(\lambda)}} \ (\lambda \in \Gamma).$$

Recall that

$$||\sum_{i=1}^{n(\lambda)} \theta_{a_i^{(\lambda)},\, x_i^{(\lambda)}}|| = ||\sum_{i=1}^{n(\lambda)} x_i^{(\lambda)} \left(a_i^{(\lambda)}\right)^*||,$$

which shows that $\sum_{i=1}^{n(\lambda)} x_i^{(n)} \left(a_i^{(\lambda)}\right)^*$ is a Cauchy net in E whose converges to y_T in E. It follows that for each $x \in E$, we have

$$T(x) = \lim_{\lambda} \sum_{i=1}^{n(\lambda)} \theta_{a_i^{(\lambda)},\, x_i^{(\lambda)}}(x) = \lim_{\lambda} \sum_{i=1}^{n(\lambda)} < x_i^{(\lambda)} \left(a_i^{(\lambda)}\right)^*,\ x >$$

$$=< \lim_{\lambda} \sum_{i=1}^{n(\lambda)} x_i^{(\lambda)} \left(a_i^{(\lambda)}\right)^*, x >=< y_T,\ x >,$$

in which we have used the continuity of A-valued inner product.

Finally, we want to prove that if y is any fixed element in E then the module mapping $T_y(x) =< y,\ x >\quad (x \in E)$ is in $K(E,\ A)$. In fact, let $\{u_\lambda\}$ be an approximate unit of A, then we have

$$T_{yu_\lambda}(x) =< yu_\lambda,\ x >= u_\lambda < y,\ x >= \theta_{u_\lambda,\ y}(x),$$

which shows that $T_{yu_\lambda} \in K(E,\ A)$. Since $||yu_\lambda - y|| \to 0,$ and since $||yu_\lambda|| = ||\theta_{u_\lambda,\ y}||$, $\{\theta_{u_\lambda,\ y}\}$ is a Cauchy net in $K(E,\ A)$ that converges to $k_y \in K(E, A)$. Therefore it follows from the limit uniquenessthat $T_y = k_y \in K(E,\ A)$ since $T_{yu_\lambda} \to T_y$.

On the other hand, in order to generalize Riesz-Frechet representation theorem, the concept of selfdual module will be introduced as follows:

From the above observation, $K(E,\ A)$ is really included in $B(E,\ A)$. If exactly $K(E,\ A) = B(E,\ A)$ holds, then E is said to be a selfdual module. At this time

$$B(E,\ A) = L(E,\ A) = K(E,\ A) \cong E.$$

From Remark 1.11 we see that unital C^*-algebra which itself as Hilbert C^*-module is selfdual. W.L. Paschke [18] proved that for any Hilbert C^*-module E over a von Neumann algebra A, $B(E,\ A)$ can become a selfdual A-module. Moreover, $L(E) = B(E)$ holds for a selfdual module E, and $L(E)$ is also a von Neumann algebra whenever A is a von Neumann algebra.

For any Hilbert C^*-module, W.L. Paschke [18] gave also a method to extend it to a selfdual module. The following is a brief introduction to this construction:

Consider algebraic tensor product $E \otimes A^{**}$, in which A^{**} is the enveloping von Neumann algebra of A, that is, the double commutant of A. $E \otimes A^{**}$ is a right A^{**}-

module by $(h \otimes a) \cdot b = h \otimes ab$ $(h \in E, a, b \in A^{**})$, then define A^{**}-valued inner product as follows:

$$[,\ ] : E \otimes A^{**} \times E \otimes A^{**} \to A,$$

$$\left[\sum_{i=1}^{n} x_i \otimes a_i,\ \sum_{j=1}^{m} y_j \otimes b_j \right] = \sum_{i=1}^{n} \sum_{j=1}^{m} a_i^* < x_i,\ y_j > b_j,$$

for all $x_i,\ y_j \in E,\ a_i,\ b_j \in A^{**},\ i = 1,\ 2,\ \cdots,\ n;\ j = 1,\ 2,\ \cdots;\ m,\ n \in \mathbb{N}$.

Let $E_0 = \{z \in E \otimes A^{**} : [z,\ z] = 0\}$, the norm closure of quotient space $E \otimes A^{**}/E_0$ induced by the above A^{**}-valued inner product is a Hilbert C^*-module over the von Neumann algebra A^{**}, which is denoted by $\overline{E}$. Furthermore, let $\widetilde{E} = B(\overline{E},\ A^{**})$, then $\widetilde{E}$ is a selfdual module, and every module mapping in $B(E)$ can be extended preserving norm to that in $B(\widetilde{E})$.

W.L. Paschke [18] also studied W^*-module. A Hilbert C^*-module E over a von Neumann algebra A is said to be a W^*-module if there is a Banach space V such that $E = V^*$. From J. Schweizer [24] we see that for any Hilbert C^*-module E over a C^*-algebra A, then E^{**} is W^*-module over von Neumann algebra A^{**}, where E^{**} is the double dual space of E as Banach space. Moreover, Hilbert C^* module E over a von Neumann algebra is selfdual if and only if E is W^*-module by Proposition 2.9 in [24].

In particular, Lin Huaxin [9], M. Frank [6] proved that $l^2(A)$ is a selfdual module if and only if A is a finite dimensional C^*-algebra, that is, the direct sum of finite matrix algebras. Further, M. Skeide [25] proved that Hilbert C^*-module E is selfdual if and only if the unit ball $S = \{x \in E : ||x|| \leq 1\}$ in E is complete with respect to the locally convex topology generated by the family of semi-norms $\{p_x(\cdot) = || < x, \cdot > || : x \in E\}$.

Remark 1.12 From the above observation we see that if E is a W^*-module then $L(E)$ is a von Neumann algebra. Thus (regular unbounded) selfadjoint module operator (via Cayley transformation) has spectral resolution, the relevant conclusions will be used in the study of quantum Markov module operators semigroups and operator valued Dirichlet forms in Chap. 3.

At the end of this subsection, we give the concept of strict topology on $L(E,\ F)$ commonly used later:

Suppose that E and F are Hilbert C^*-modules over a C^*-algebra A, then the strict topology on $L(E,\ F)$ is defined to be the topology given by the seminorms

$$T \to ||Tx||; T \to ||T^*y||\ for\ all\ x \in E,\ y \in F.$$

When E and F degenerate to Hilbert space H, the strict topology defined above is the strong $*$ topology in $B(H)$. It is well known that the unit ball of $K(H)$ is strong $*$ topological dense in $B(H)$. Similarly, the following conclusion holds for Hilbert C^*-modules:

Proposition 1.6 *Suppose that E and F are Hilbert C^*-modules over a C^*-algebra A. Then the unit ball of $K(E, F)$ is strictly dense in the unit ball of $L(E, F)$.*

Proof We need to show that for any $T \in L(E, F)$ and $||T|| \leq 1$, then there exist a net $\{T_\lambda\} \subset K(E, F)$ such that

$$||T_\lambda x - Tx|| \to 0 (x \in E),$$

and

$$||T_\lambda^* y - T^* y|| \to 0 \ (y \in F).$$

For this, let $\{u_\lambda\}$ be an approximate unit of $K(E)$, and put $T_\lambda = Tu_\lambda$. Since

$$L(E, F)K(E) \subset K(E, F); \ K(F)L(E, F) \subset K(E, F),$$

this shows that $T_\lambda \in K(E, F)$. We next need only to consider the elements in $E < E, E >$ because $E < E, E >$ is dense in E, Since $||\theta_{x, y} - u_\lambda \theta_{x, y}|| \to 0 \ (x, y \in E)$, we have

$$||T_\lambda x < y, w > -Tx < y, w > ||$$

$$= ||Tu_\lambda \theta_{x, y}(w) - T\theta_{x, y}(w)||$$

$$\leq ||\theta_{x, y}(w) - u_\lambda \theta_{x, y}(w)||$$

$$\leq ||\theta_{x, y} - u_\lambda \theta_{x, y}|| \cdot ||w|| \to 0 \ (x, y, w \in E).$$

Furthermore, since

$$||x < y, w > -u_\lambda x < y, w > ||$$

$$= ||\theta_{x, y}(w) - u_\lambda \theta_{x, y}(w)|| \to 0,$$

it follows that $||x - u_\lambda x|| \to 0$ for all $x \in E$. Hence

$$||T_\lambda^* y - T^* y|| = ||u_\lambda T^* y - T^* y|| \to 0,$$

for all $y \in F$ since $T^* y \in E$.

1.1.3 Multiplier Theorem

In the study of C^*-algebras, the case with unit element is easier to handle than that without unit, so how to add unit element becomes an important problem. If a C^*-algebra A is not unital, then there are generally three ways to add unit as follows:

(1) $A^+ = A \oplus \mathbb{C}$;
(2) multiplier algebra $M(A) = \{x \in A^{**} : \ xA \subset A, \ Ax \subset A\}$;
(3) the double commutant $A^{''} = A^{**}$ of A.

From the above constructions we see that A is a C^*-subalgebra of the above three algebras. Meanwhile, it is a closed two-sided ideal of A^+ and $M(A)$, respectively. But in general, it is not the ideal of A^{**}. In particular, if A has unit then $M(A) = A$.

In order to describe multiplier theorem, first we introduce briefly the basic knowledge of multiplier algebra, which is of great important in more advanced aspects of the C^*-algebra theory, especially in certain approaches to K-theory.

Suppose that A is a C^*-algebra, the multiplier on A is a pair of $(L, \ R)$ which satisfies the following relation:

$$R(a)b = aL(b) \ (a, \ b \in A),$$

in which $L \in B(A)$ is right module map (namely, $L(ab) = L(a)b$), $R \in B(A)$ is left module map (namely, $R(ab) = aR(b)$), where $B(A)$ is denoted by the Banach space consisting of all bounded linear maps from A to A. We write $M(A)$ to be the set of all multipliers on A. If $(L, \ R) \in M(A)$, then define

$$||L|| = sup_{||a|| \leq 1} ||L(a)||.$$

Similarly, $||R||$ can be defined and it is easily checked that $||L|| = ||R||$. Thus we can introduce a norm on $M(A)$ by $||(L, \ R)|| = ||L||$, hence $M(A)$ is a closed subspace of $B(A) \oplus B(A)$ with respect to the above norm.

Next, define a multiplication operation on $M(A)$ by the following way:

$$(L_1, \ R_1) \cdot (L_2, \ R_2) = (L_1 L_2, \ R_2 R_1),$$

for all $(L_1, \ R_1), \ (L_2, \ R_2) \in M(A)$. Straightforward computation shows that the above product $(L_1 L_2, \ R_2 R_1)$ is in $M(A)$ and that $M(A)$ is an algebra under this multiplication.

We now introduce the $*$ operation in $M(A)$: If $(L, \ R) \in M(A)$ then define

$$L^*(a) = (R(a^*))^* \ (a \in A),$$

and

$$R^*(a) = (L(a^*))^* \ (a \in A).$$

It is easy to see that $(L^*, R^*) \in M(A)$, and $* : (L, R) \to (L^*, R^*)$ is an involution operation. It is further proved that $M(A)$ becomes a C^*-algebra with respect to the above operations and contains A as maximal essential ideal. Details can refer to Section 2.1 of Chapter 2 in [17], or Chapter 2 in [13].

Remark 1.13 It can be proved that the above two definitions of multiplier algebra are equivalent.

Suppose that A is a C^*-algebra, we now consider the corresponding relationship between $L(A)$ and $M(A)$. Let $\{u_\lambda\}$ be an approximate unit of A, and let $L \in L(A)$, define

$$R(a) = \lim_\lambda a\ L(u_\lambda)\ (a \in A),$$

it is easy to check that (L, R) is a multiplier on A, namely, $(L, R) \in M(A)$. Then define a linear map $\Phi : L(A) \to M(A)$ given by

$$L \to (L, R).$$

Since $||\Phi(L)|| = ||(L, R)|| = ||L||$, Φ is an isometry. By routine calculation we can get

$$\Phi(L_1 L_2) = \Phi(L_1)\Phi(L_2)\ (L_1, L_2 \in L(A));$$

$$\Phi(L^*) = (\Phi(L))^*\ (L \in L(A)),$$

which shows that Φ is an isometric $*$ isomorphism.

Finally, to see that Φ is onto. For any $(L, R) \in M(A)$, since $R(a) = \lim_\lambda R(a)u_\lambda = \lim_\lambda aL(u_\lambda)$, and L has adjoint L^* given by $L^*(a) = (R(a^*))^*$ $(a \in A)$, it follows that L is in $L(A)$ and $\Phi(L) = (L, R)$. The above proof shows that $L(A) \cong M(A)$.

Remark 1.14

(1) Combining Example 1.5 with the above conclusion, we have

$$L(A) \cong M(A) \cong M(K(A));$$

(2) Let H be a separable Hilbert space. It is well known that $K(H)$ is a maximal essential ideal of $B(H)$, and we have then

$$B(H) \cong M(K(H)) = \{T \in B(H) :\ TK(H) \subset K(H),\ K(H)T \subset K(H)\}.$$

Based on the above observation, K. Kasprov [11] extended the above conclusion to the general Hilbert C^*-modules and obtained the following conclusion:

Theorem 1.1 (Multiplier Theorem) *Let E be a Hilbert C^*-module over a C^*-algebra A, then we have*

$$L(E) \cong M(K(E)).$$

Proof Suppose that $T \in L(E)$. Define A-linear mappings L_T and R_T which from $K(E)$ to itself by

$$L_T S = TS \ (S \in K(E)),$$

and

$$R_T S = ST \ \ (S \in K(E)),$$

respectively. It is easy to check that $(L_T,\ R_T)$ is in $M(K(E))$. Then we can define a A-linear mapping $\Phi:\ L(E) \to M(K(E))$ given by

$$T \to (L_T,\ R_T)\ (T \in L(E)).$$

Similar to the above proof of $L(A) \cong M(K(A))$, we have that Φ is an isometric $*$ isomorphism. Next, to show that Φ is onto. Let $\{u_\lambda\}$ be an approximate unit of $K(E)$. For any $(L,\ R) \in M(K(E))$, define A-linear mappings T_L and T_L^* which from E to itself by

$$T_L(x) = \lim_\lambda L(u_\lambda)x \ (x \in E),$$

and

$$T_L^*(x) = \lim_\lambda (L(u_\lambda))^* x \ (x \in E),$$

respectively. Then we have

$$< T_L(x),\ y > = < \lim_\lambda L(u_\lambda)x,\ y >$$

$$= \lim_\lambda < L(u_\lambda)x,\ y > = \lim_\lambda < x,\ (L(u_\lambda))^* y >$$

$$= < x,\ \lim_\lambda (L(u_\lambda))^* y > = < x,\ T_L^* y > \ \ (y \in E),$$

which shows that T_L is in $L(E)$.

Finally, we have to prove that $L = L_{T_L}$ and $R = R_{T_L}$. Since if it is proved, $\Phi(T_L) = (L_{T_L},\ R_{T_L}) = (L,\ R)$, so that Φ is onto. In fact, we need only to

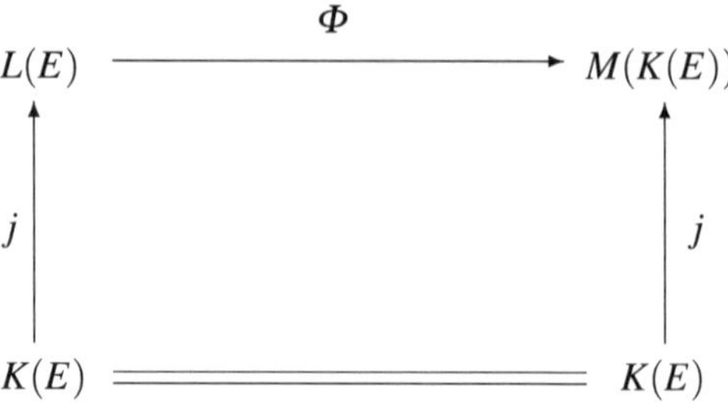

Fig. 1.1 Exchange chart

calculate whether they are true on $\theta_{x,\ y},\ \ (x,y\in E)$. Since $T_L(x)=\lim_\lambda L(u_\lambda)x$ for $x\in E$, we have

$$L_{T_L}(\theta_{x,\ y})=T_L\cdot\theta_{x,\ y}$$

$$=\lim_\lambda L(u_\lambda)\theta_{x,\ y}=L(\lim_\lambda u_\lambda\theta_{x,\ y})$$

$$=L(\theta_{x,\ y}).$$

Hence $L_{T_L}=L$. From the definition of multiplier we see that L and R are uniquely determined each other, so $R_{T_L}=R$, the desired claim is proved.

Refer to the exchange chart as follows (Fig. 1.1):
where j is the embedding mapping. From Theorem 1.1 we see that $K(E)$ is an essential ideal of $L(E)$, so Φ is the unique extension of j. Moreover, we have

$$M(K(E))=\{T\in L(E):T\ K(E)\subset K(E);\ \ K(E)\ T\subset K(E)\},$$

in the sense of isometric isomorphism,

Remark 1.15 The above proof indicates that $M(K(E))$ is the closure of $K(E)$ with respect to the strict topology. In addition, Huaxin Lin [10] established corresponding isometric isomorphism relations in a wider range of module mappings as follows:

$$B(E)\cong LM(K(E));\ \ B(E,\ B(E,\ A))\cong QM(K(E)),$$

where $LM(A)=\{x\in A^{**}:x\ a\in A,\ \ a\in A\}$ is the set of all left multipliers on C^*-algebra A, and $QM(A)=\{x\in A^{**}:a\ x\ b\in A,\ a,\ b\in A\}$ is the set of all quasi-multipliers on A. In particular, if E is selfdual then we have

$$M(K(E))=LM(K(E))=QM(K(E))=B(E).$$

Example 1.6 If Φ is a $*$ isomorphism from C^*-algebra A onto C^*-algebra B, then there is a unique $*$ isomorphism $\overline{\Phi}$ from $M(A)$ onto $M(B)$ whose restriction to A is Φ..

Indeed, since $A \overset{\Phi}{\rightarrow} B \overset{j}{\rightarrow} M(B)$, in which j is an embedding mapping, the composition mapping $j \circ \Phi : A \rightarrow M(B)$ is injective. It follows that there exists a uniquely $*$ isomorphism $\overline{\Phi}$ from $M(A)$ to $M(B)$ whose restriction to A is Φ by Theorem 1.1. We prove next that $\overline{\Phi}$ is surjective. From $B \overset{\Phi^{-1}}{\rightarrow} A \overset{j}{\rightarrow} M(A)$, we see that $j \circ \Phi^{-1} : B \rightarrow M(A)$ is injective. Therefore there is a uniquely $*$ isomorphism $\overline{\Psi}$ from $M(B)$ to $M(A)$ whose restriction to B is $j \circ \Phi^{-1}$ by Theorem 1.1 again. It follows from the above proof that $\overline{\Phi} \circ \overline{\Psi} : M(B) \rightarrow M(B)$ is a $*$ isomorphism extension of $j : B \rightarrow M(B)$. However, the identity mapping I also meets the requirements. Therefore it can be obtained by the uniqueness of expansion

$$\overline{\Phi} \circ \overline{\Psi} = I.$$

Similarly

$$\overline{\Psi} \circ \overline{\Phi} = I,$$

which shows that $\overline{\Phi}$ is onto, where I is the identity mapping.

1.2 Polar Decomposition and Wold Decomposition

1.2.1 Unitary Equivalence Between Hilbert C-Modules*

We first recall the concept of unitary equivalence between Hilbert spaces. Suppose that H and K are Hilbert spaces. We say that they are unitary equivalent if there is a linear bijection $U : H \rightarrow K$ such that

$$< Ux,\ Uy >=< x,\ y > (\ x,\ y \in H),$$

in this case, the above operator U is said to be unitary. Moreover, we have then $< x,\ U^{-1}y >=< Ux,\ y >$, which shows that $U^* = U^{-1}$, that is, $U^*U = I_H;\ UU^* = I_K \quad (I)$, where I_H and I_K are the identity operators on H and K, respectively. The above process is reversible. Therefore linear operators satisfying formula (I) are usually used as the definition of unitary operators.

Similarly, in Hilbert C^*-module theory, the concepts of unitary module operator and unitary equivalence will be introduced as follows.

Definition 1.7 Let E and F be Hilbert C^*-modules over a C^*-algebra A. If there exists a bijective A-linear map U from E onto F which satisfies the following equation

$$< Ux, Uy >=< x, y > (x, y \in E),$$

then we say that E and F are unitary equivalent, which is denoted by $E \cong F$.

Similar to the above process in the case of Hilbert spaces, we have then

$$U^*U = I_E;\ UU^* = I_F.\ (II)$$

The module operator satisfying formula (II) is called unitary module operator.

It is well known that for Hilbert spaces H and K, linear operators $U : H \to K$, is unitary if and only if U is isometric surjective, that is, $||Ux|| = ||x||$ for every x in H.

Indeed, since $||Ux|| = ||x|| \Leftrightarrow < Ux,\ Uy > = < x, y >$ for all $x,\ y \in H$. Hence the necessary is proved by letting $y = x$. As for the proof of sufficiency, it can be done by the polarization identity. E.C. Lance [14] extended the above result to Hilbert C^*-modules and obtained the following conclusion:

Theorem 1.2 *Let E and F be Hilbert C^*-modules over a C^*-algebra A. Then A-linear mapping U from E to F is a unitary module operator if and only if U is an isometric surjection*

The proof of Theorem 1.2 is much more complicated than that of Hilbert space. First a lemma is given.

Lemma 1.1 *Let A be a C^*-algebra. If positive elements a and b in A satisfy the equation $|| < a,\ c > || = || < b,\ c > ||$ for all $c \in A$, then $a = b$. Where A is regarded itself as Hilbert C^*-module with inner product $< a, b > = a^*b$.*

Proof First, we have $||a|| = ||b||$ since $|| < a,\ c > || = || < b,\ c > ||$ for all $c \in A$. Hence, without loss of generality, in the following assume that $||a|| = ||b|| = 1$. By function calculus of positive elements of C^*-algebras, to see $a = b$ it suffices to prove $a^2 = b^2$ since $a = (a^2)^{\frac{1}{2}} = (b^2)^{\frac{1}{2}} = b$.

Assume that $a^2 - b^2 \neq 0$. Since $a^2 - b^2$ is a selfadjoint element in A, the spectral set $\sigma(a^2 - b^2)$ contains at least $||a^2 - b^2||$ or $-||a^2 - b^2||$, so $\sigma(a^2 - b^2) \neq \{0\}$. Suppose that $\sigma(a^2 - b^2)$ contains $||a^2 - b^2||$, because otherwise $\sigma(b^2 - a^2)$ can be considered. Since the spectral radius $r(a^2 - b^2) \leq ||a^2 - b^2||$, it follows that

$$||a^2 - b^2|| = \max_{t \in \sigma(a^2 - b^2)} \{t\}.$$

Construct a real valued continuous function defined on $\sigma(a^2 - b^2)$ by the following way

$$f(t) = \begin{cases} 0, & \text{if } t \in \sigma(a^2 - b^2) \cap \left[-||a^2 - b^2||,\ \frac{1}{6}||a^2 - b^2||\right], \\ \frac{t - \frac{1}{6}||a^2 - b^2||}{\frac{5}{6}||a^2 - b^2||}, & \text{if } t \in \sigma(a^2 - b^2) \cap \left(\frac{1}{6}||a^2 - b^2||,\ ||a^2 - b^2||\right]. \end{cases}$$

From the function calculus of selfadjoint elements of C^*-algebra, it is easy to see that $c = f(a^2 - b^2) \in A$ and we have

$$||c(a^2 - b^2)c|| = \max_{t \in \sigma(a^2 - b^2)} |f(t)tf(t)| = ||a^2 - b^2||.$$

Since $f(t)tf(t) \geq \frac{1}{6}||a^2 - b^2||f^2(t)$, by the function calculus of selfadjoint elements again we have then

$$c(a^2 - b^2)c \geq \frac{1}{6}||a^2 - b^2||c^2 \tag{1.1}$$

Choose a state ρ on A such that $\rho(cb^2c) = ||cb^2c||$. Note that $||b|| = 1$ and b is positive, then we have $b^2 \leq b \leq 1$ and $cb^2c \leq c^2$. from which implies that $\rho(cb^2c) \leq \rho(c^2)$. By the above formula (1.1) again one can obtain

$$ca^2c \geq cb^2c + \frac{1}{6}||a^2 - b^2||c^2.$$

Hence

$$\rho(ca^2c) \geq \rho\left(cb^2c + \frac{1}{6}||a^2 - b^2||c^2\right)$$

$$= \rho(cb^2c) + \frac{1}{6}||a^2 - b^2||\rho(c^2)$$

$$\geq \left(1 + \frac{1}{6}||a^2 - b^2||\right)\rho(cb^2c),$$

which implies that

$$||ca^2c|| \geq \rho(ca^2c)$$

$$\geq \left(1 + \frac{1}{6}||a^2 - b^2||\right)||cb^2c||.$$

The above proof shows that if $cb^2c \neq 0$ then $||ca^2c|| > ||cb^2c||$. In addition, the case of $cb^2c = 0$ is clear since $ca^2c > 0$ from the above formula (1.1). So in either case we have $||ca^2c|| > ||cb^2c||$, namely, $|| < a, c >^* < a, c > || > || < b, c >^* < b, c > ||$, equivalently, $|| < a, c > || > || < b, c > ||$. This contradicts the assumption of the lemma.

***The proof of Theorem* 1.2** Suppose that U is unitary. Then we have

$$||Ux||^2 = || < Ux, Ux > || = || < U^*Ux, x > ||$$

$$= || < x, x > || = ||x||^2 \; for \; all \; x \; in \; E.$$

Hence U is isometry. Since $UU^* = I_F$, U is surjective, where I_F is denoted the identity mapping on F.

Conversely, if U is an isometric surjection, then it follows for any x in E and any a in A that

$$||a^* < Ux,\ Ux > a|| = || < U(xa),\ U(xa) > ||$$

$$= ||U(xa)||^2 = ||x||^2 = || < xa,\ xa > ||$$

$$= ||a^* < x,\ x > a||.$$

Therefore

$$|| << Ux,\ Ux >^{1/2},\ a > || = || << x,\ x >^{1/2},\ a > ||,$$

since $||a^*a|| = ||a||^2$. It results from Lemma 1.1 that

$$< Ux,\ Ux >^{1/2} = < x,\ x >^{1/2}.$$

Hence by the polarization identity, we have

$$< Ux,\ Uy > = \frac{1}{4}\{< U(x+y),\ U(x+y) > - < U(x-y),\ U(x-y) >$$

$$+ i < U(ix+y),\ U(ix+y) > -i < U(ix-y),\ U(ix-y) >\}$$

$$= \frac{1}{4}\{< x+y,\ x+y > - < x-y,\ x-y >$$

$$+ i < ix+y,\ ix+y > -i < ix-y,\ ix-y >\}$$

$$= < x,\ y >,$$

for all $x,\ y \in E$. Thus $< Ux,\ y > = < x,\ U^{-1}y >$, which shows that $U^* = U^{-1}$. Hence U is unitary.

Remark 1.16

(1) In fact, Lemma 1.1 is a special case of Theorem 1.2. For any given positive elements a and b in A, denote by $[aA]$ and $[bA]$ the closed right ideals which are the norm closures of aA and bA in A, respectively. Hence, they are Hilbert C^*-modules over A, respectively. Define a linear mapping U from aA onto bA by

$$U : ax \to bx,$$

for all $x \in A$. Since $||ax|| = ||bx||$, U can be extended to an isometric surjection from $[aA]$ onto $[bA]$.

(2) Theorem 1.2 indicates that if U is isometric surjective, then the adjoint operator U^* exists, and $U^*U = I_E$; $UU^* = I_F$. But if U is only an isometric (not necessarily onto) module operator, then U^* does not necessarily exist. The counterexample of Sect. 1.1.2, $j : K(H) \to B(H)$ fully illustrates this point.

Further observation shows that the reason why the embedded mapping j has no adjoint module operator is that $j(K(H)) = K(H)$ is not complemented in $B(H)$. If the complemented condition is added, the characterization of isometric module operators can be obtained which is similar to the case of Hilbert space:

Proposition 1.7 *Suppose that E and F are Hilbert C^*-modules over a C^*-algebra A, and $T : E \to F$ is a A-linear mapping. Then the following statements are equivalent:*

(1) $T \in L(E, F)$ and $T^*T = I_E$, where I_E is the identity mapping on E;
(2) $||Tx|| = ||x||$ $(x \in E)$ and $R(T) = \{Tx : x \in E\}$ is complemented in F.

Proof (1) $\Rightarrow$ (2) : Since $T^*T = I_E$, we have then

$$||Tx||^2 = || < Tx, Tx > ||$$

$$= || < T^*Tx, x > || = || < x, x > ||$$

$$= ||x||^2,$$

For all $x \in E$. So $||Tx|| = ||x||$ for all $x \in E$. Since

$$(TT^*)^2 = (TT^*)(TT^*)$$

$$= T(T^*T)T^* = TT^*,$$

TT^* is a projection element in $L(F)$. Observe that $(TT^*)T = T(T^*T) = T$, hence $R(TT^*) = R(T)$, where $R(TT^*)$ and $R(T)$ stand for the range of TT^* and T, respectively. For each $y \in F$, from the above proof, we have the following orthogonal decomposition

$$y = TT^*y + (I_F - TT^*)y,$$

this shows that $F = R(T) \oplus R(I_F - TT^*)$. Hence $R(T) = R(TT^*)$ is complemented.

(2) $\Rightarrow$ (1) : Since $T : E \to R(T)$ is an isometric surjection, therefore from Theorem 1.2 that there exists module mapping $\overline{T^*} : R(T) \to E$ such that $\overline{T^*}T = I_E$. Define a module mapping $T^* : F \to E$ given by the following way

$$T^*(y) = \begin{cases} \overline{T^*}y, & \text{if } y \in R(T), \\ 0, & \text{if } y \in R(T)^{\perp}. \end{cases}$$

Then it is easy to check that $T^* \in L(F, E)$ since $R(T) \oplus R(T)^{\perp} = F$. Routine calculation shows that T^* is the adjoint operator of T. Thus $T \in L(E, F)$, and from the construction of T^* we have $T^*T = \overline{T^*}T = I_E$.

Example 1.7 If given a Hilbert space H and a compact Hausdorff space X. Then $H \otimes C(X)$ and $C(X, H)$ are unitary equivalent as Hilbert C^*- modules over C^*-algebra $C(X)$, namely,

$$H \otimes C(X) \cong C(X, H).$$

In fact, let $U : \xi \otimes f \to U_{\xi \otimes f} \in C(X, H)$, where $U_{\xi \otimes f}(x) = f(x)\xi$ for all $x \in X$ and $f \in C(X)$. It is easy to check that U is unitary.

Many important results of operator theory on Hilbert spaces can be extended to the case of Hilbert C^*-modules if the complemented condition is added. The following proposition that describes complementarity of kernel space and range space of module operator is often used in the rest of this book.

Proposition 1.8 (see [13], Theorem 3.2) *Suppose that E and F are Hilbert C^*-modules over a C^*-algebra A. If $T \in L(E, F)$ has close range $R(T)$, then T^* has closed range $R(T^*)$ also, and $ker(T)$, $R(T)$ are complemented in E and F respectively, where $kerT = \{x \in E : Tx = 0\}$ is the kernel of T. Meanwhile, the same conclusion holds for $ker(T^*)$ and $R(T^*)$.*

Proof First, we prove the case when T is surjective. Let E_1 be the closed unit ball of E. Therefore, according to the open mapping theorem, there exists a $\lambda > 0$ and closed ball $B(0, \lambda) \subset F$ centered at the origin and of radius λ such that $B(0, \lambda) \subset T(E_1)$, that is,

$$B(0, \lambda) = \{y \in F | ||y|| \leq \lambda\} \subset T(E_1) = \{T(x) | x \in E, ||x|| \leq 1\}.$$

Hence for any $y (\neq 0) \in F$, there exists $\tilde{x} \in E_1$ such that $T(\tilde{x}) = \frac{\lambda}{||y||} y$ since $\frac{\lambda}{||y||} y \in B(0, \lambda)$. It follows that $T(\frac{||y||}{\lambda} \tilde{x}) = y$. Let $x = \frac{||y||}{\lambda} \tilde{x}$. Since $||\tilde{x}|| \leq 1$, $||x|| \leq ||y|| \lambda^{-1}$ and $Tx = y$ by the above proof. Since

$$||T^* y||^2 = || < y, TT^* y > || \leq ||y|| \cdot ||TT^* y||,$$

from which it implies that

$$||y||^2 = || < Tx, y > || = || < x, T^* y > ||$$

$$\leq ||x|| ||T^* y|| \leq \lambda^{-1} ||y|| \cdot ||y||^{1/2} ||TT^* y||^{1/2}$$

$$= \lambda^{-1} ||y||^{\frac{3}{2}} ||TT^* y||^{\frac{1}{2}}.$$

Thus $||y|| \leq \lambda^{-2}||TT^*y||$, this inequality holds for $y = 0$ as well. Hence TT^* in $L(F)$ is invertible and surjective by [13], Lemma 3.1. Therefore for any $z \in E$, there exists a $w \in F$ such that $Tz = TT^*w$, this shows that $z - T^*w \in kerT$. Hence

$$z = (z - T^*w) + T^*w \in kerT + R(T^*).$$

It is easy to prove that

$$kerT = R(T^*)^{\perp}; \ \ (kerT)^{\perp} = R(T^*).$$

The above proof shows that T^* has closed range $R(T^*)$, and $kerT$ and $R(T^*)$ are complemented in E, respectively. By exchanging T and T^*, similar to the above proof process, we can obtain

$$kerT^* = R(T)^{\perp}; \ \ (kerT^*)^{\perp} = R(T),$$

and

$$R(T) \oplus R(T)^{\perp} = F.$$

Hence $kerT^*$ and $R(T)$ are complemented in F, respectively.

We next consider the case when $R(T)$ is not equal to F. Write $R(T)$ as F_0 and rewrite T as $\hat{T}$,, then it is easy to prove that $\hat{T}^* = T^*|_{F_0} \in L(F_0, \ E)$. In what follows, replace F with F_0, and repeat the above proof process, then we have

$$(kerT)^{\perp} = (ker\hat{T})^{\perp} = R(\hat{T}^*).$$

Also, it is easy to prove that $R(T^*) \subset (kerT)^{\perp}$. Therefore $R(T^*) \subset R(\hat{T}^*)$. On the other hand, since $T^*|_{F_0} = \hat{T}^*$, $R(\hat{T}^*) \subset R(T^*)$. It follows that $R(\hat{T}^*) = R(T^*)$. So the case of $R(T) \neq F$ can be transformed into the case of the previous case of $R(T) = F$. So we are done.

From Proposition 1.8, we immediately get the following consequence:

Corollary 1.3 *If E is a Hilbert A-modules and F is a closed submodule of E. Then F is complemented in E if and only if there exists $T \in L(E)$ such that $R(T) = F$.*

Remark 1.17 It is easy to check that the following relations hold for all $T \in L(E, \ F)$:

$$R(T) \subseteq (kerT^*)^{\perp} \text{ and } R(T)^{\perp} = kerT^*.$$

But in general, $(kerT^*)^{\perp} \neq \overline{R(T)}$.

Magajna [16] proved that every nontrivial closed submodule in a full Hilbert C^*-module E over a C^*-algebra A is complemented if and only if A is $*$ isomorphic to a C^*-subalgebra of $K(H)$ (which is called dual algebra), where H is a Hilbert space.

Inspired by the work of Magajna [16], in [31], we proved that for a given Hilbert C^*-module E, then every closed submodule of E is complemented if and only if every hereditary subalgebra B of $K(E)$ has the form $B = pK(E)p$, where p is a projection element in $L(E)$. Here we remove the restriction that Hilbert C^*-module is full. In general, Hilbert C^*-modules are not full, such as the proper closed right ideals of C^*-algebras.

1.2.2 Polar Decomposition and Wold Decomposition

Definition 1.8 Let E and F be Hilbert C^*-modules over a C^*-algebra A, and let $T \in L(E, F)$. If $R(T)$ is closed and $||Tx|| = ||x||$ for all $x \in (kerT)^{\perp}$, then T is called a partial isometric module operator.

Remark 1.18 In Definition 1.8, if $kerT = \{0\}$, then T is the isometric module operator mentioned above. Furthermore, if T is surjective also, then it is unitary by Theorem 1.2.

Similar to the case of Hilbert space, some equivalent characterizations of partial isometric module operator are given as follows:

Suppose that $T \in L(E, F)$. Then the following conditions are equivalent:

(1) T partial isometric module operator;
(2) T^*T is a projection in $L(E)$;
(3) TT^* is a projection in $L(F)$;
(4) $TT^*T = T$;
(5) $T^*TT^* = T^*$.

The above equivalence characterizations can be easily obtained by Definition 1.8.

It is known that if operator T is in $B(H)$, where H is a Hilbert space, then T has polar decomposition, that is, $T = V(T^*T)^{1/2}$, in which V is a partial isometric operator uniquely determined by T and satisfies $kerT = ker((T^*T)^{1/2}) = kerV$.

This conclusion can be extended to the case of von Neumann algebra. However, for general C^*-algebra A, the partial isometric operator occurring in polar decomposition belongs to A^{**} and generally does not belong to A.

If the complemented condition is added, then the following conclusion can be obtained for Hilbert C^*-modules:

Proposition 1.9 *Suppose that E is a Hilbert A-module, and suppose that $T \in L(E)$ has closed range $R(T)$. Then T has polar decomposition $T = V(T^*T)^{1/2}$, where V is the only partial isometric module operator in $L(E)$ satisfying $kerV = kerT$.*

Proof Since $R(T)$ is closed, $R(T^*)$ is closed also by Proposition 1.8, we have

$$E = kerT^* \oplus R(T) = kerT \oplus R(T^*).$$

It implies that $R(T^*T) = R(T^*)$.

Also

$$(T^*T)E = (T^*T)^{1/2} \cdot (T^*T)^{1/2}E$$

$$\subset (T^*T)^{1/2}E \subset (T^*T)E.$$

So that $R((T^*T)^{1/2}) = R(T^*T)$. From the above proof, we can get $R((T^*T)^{1/2}) = R(T^*)$. Since

$$||(T^*T)^{1/2}x||^2 = || < (T^*T)^{1/2}x, (T^*T)^{1/2}x > ||$$

$$= || < (T^*T)x, x > || = || < Tx, Tx > ||$$

$$= ||Tx||^2,$$

for all $x \in E$. Hence, the following defined A-linear mapping

$$V_0 : (T^*T)^{1/2}(E) \to T(E),\ (T^*T)^{1/2}x \to Tx\ (x \in E)$$

is isometric. Next, define a module mapping as follows:

$$V : E \to E, V(x) = \begin{cases} V_0^{-1}(x), & \text{if } x \in R(T^*), \\ 0, & \text{if } x \in R(T^*)^{\perp}. \end{cases}$$

By routine calculation, the adjoint of V is that

$$V^* : E \to E, V^*(x) = \begin{cases} V_0^{-1}(x), & \text{if } x \in R(T), \\ 0, & \text{if } x \in R(T)^{\perp}. \end{cases}$$

Therefore $V \in L(E)$,, and from the constructions of V and V^*, we have that $T = V(T^*T)^{1/2}$, and $kerV = R(T^*)^{\perp} = kerT$; $VV^* = P_{R(T)}$, where $P_{R(T)}$ is a projection from E onto $R(T)$. Hence V a is partial isometric module operator.

Finally, we have to prove uniqueness of polar decomposition. Assume that there is another $V' \in L(E)$ such that $T = V'(T^*T)^{1/2}$, and $kerV' = kerT$. Hence

$$V'|_{R((T^*T)^{1/2})} = V|_{R((T^*T)^{1/2})},$$

and

$$V'|_{kerT} = 0 = V|_{kerT}.$$

Since

$$R((T^*T)^{1/2}) \oplus kerT$$

$$= R(T^*) \oplus kerT = E,$$

from which it shows that $V' = V$.

Remark 1.19

(1) In the proof of Proposition 1.9, $(T^*T)^{1/2}E \subset T^*TE$ has been used. This is because according to the Weierstrass approximation theorem, there exists a real coefficient polynomial $\{p_n(t)\}$ with constant term 0 converging uniformly to $t^{\frac{1}{2}}$ on $\sigma(T^*T)$. Then by the function calculus of positive elements of C^*-algebra, we have

$$(T^*T)^{1/2} = \lim_{n\to\infty} p_n(T^*T),$$

in the sense of norm topology. Hence

$$(T^*T)^{1/2}E = \lim_{n\to\infty} p_n(T^*T)E \subset T^*TE.$$

(2) Under the condition of Proposition 1.9, we can get $V^*T = (T^*T)^{1/2}$;
(3) If $T \in L(E)$ is an invertible module operator, then V in the polar decomposition is a unitary operator.

Corollary 1.4 *Suppose that E and F are Hilbert A-module. If* $T \in L(E,\ F)$ *and* $\overline{R(T)}$ *and* $\overline{R(T^*)}$ *are complemented in F and E, respectively, then* $T = V(T^*T)$, *where* $V \in L(E,\ F)$ *is a partial isometric module operator.*

In order to characterize the factorization of bounded module mappings, in [30], we extended the concept of Moore-Penrose generalized inverse to Hilbert C^*-module, that is,

Definition 1.9 Let E and F be Hilbert C^*-module over a C^*-algebra A, and let T be in $L(E,\ F)$. If there exists a $T^+ \in L(F,\ E)$ and the following conditions are satisfied:

(1) $TT^+T = T$;
(2) $T^+TT^+ = T^+$;
(3) $(TT^+)^* = TT^+$;
(4) $(T^+T)^* = T^+T$.

Then T^+ is said to be the Moore-Penrose generalized inverse module mapping of T.

Remark 1.20

(1) In particular, if the base space C^*-algebra A reduces to the complex field $\mathbb{C}$, then T^+ in Definition 1.9 becomes the classical Moore-Penrose generalized inverse on Hilbert space;
(2) It is easy to prove that T^+ satisfying the Definition 1.9 is unique and has the following basic properties:

$$R(T^+T) = R(T^*),$$

and

$$(T^+)^* = (T^*)^+.$$

Theorem 1.3 (see [32]) *Let E and F be Hilbert C^*-module over a C^*-algebra A and $T \in L(E,\ F)$. Then the following statements are equivalent:*

(1) T has Moore-Penrose generalized inverse module mapping $T^+ \in L(F,\ E)$;
(2) $R(T)$ is closed in F.

Proof (1) $\Rightarrow$ (2) : Put $P_T = TT^+$ and $P_{T^+} = T^+T$. From the meaning of T^+ we have

$$P_T^2 = P_T = P_T^*,$$

and

$$P_{T^+}^2 = P_{T^+} = P_{T^+}^*,$$

which shows that P_T and P_{T^+} are projections in $L(F)$ and $L(E)$, respectively. Hence $R(P_T)$ is closed in F and $R(P_T) = R(TT^+) \subset R(T)$. Since $P_TT = TT^+T = T$, $R(T) \subset R(P_T)$, and $R(T) = R(P_T)$. Therefore $R(T)$ is a closed submodule in F.

(2) $\Rightarrow$ (1) : Since $R(T)$ is closed, by Proposition 1.8 $R(T^*)$ is closed and

$$E = kerT \oplus R(T^*),$$

and

$$F = ker(T^*) \oplus R(T).$$

Define

$$T^+ : F \to E,\ T^+x = \begin{cases} (T|_{ker(T)^{\perp}})^{-1}x, & \text{if } x \in R(T), \\ 0, & \text{if } x \in kerT^*. \end{cases}$$

and define

$$(T^+)^* : E \to F,\ (T^+)^*(x) = \begin{cases} (T^*|_{ker(T^*)^\perp})^{-1}x, & \text{if } x \in R(T^*), \\ 0, & \text{if } x \in kerT. \end{cases}$$

Routine calculation shows that $(T^+)^*$ defined above is indeed the adjoint of the above T^+. So $T^+ \in L(F,\ E)$ and $(T^+)^* \in L(E,\ F)$.

Further, by the constructions of T^+ and $(T^+)^*$, it is easy to check that T^+ satisfies the conditions of Definition 1.9. Hence T^+ is the Moore-Penrose inverse module mapping of T.

When A is a unital C^*-algebra. We regard it itself as a Hilbert C^*-module and then identify A with $L(A)$ in the sense of isometric isomorphism, similar to Definition 1.9. The the concept of generalized inverse element can be introduced in A. The main result of R. Harte and M. Mbekhta [8] on the generalized inverse elements of C^*-algebras is the direct consequence of Theorem 1.3.

Corollary 1.5 *Suppose that A is a unital C^*-algebra and $a(\neq 0) \in A$. Then a has Moore-Penrose generalized inverse element in A if and only if aA is a closed right ideal in A.*

Indeed, let $E = F = A$ in Theorem 1.3. Since A is unital, every element in A can be represented as a left multiplication operator. Therefore by Theorem 1.2, the following A-linear given by

$$U : A \to L(A),\ a \to L_a,\ (a \in A),$$

is unitary, where $L_a(b) = ab,\ (b \in A)$ is left multiplication operator. Thus we are done by Theorem 1.3.

In 1966, R.G. Douglas [4] obtained the factorization formula for a pair of bounded linear operators with range inclusion relation on Hilbert space. In what follows, we shall extend the R.G. Douglas' factorization theorem to Hilbert C^*-module by using Moore-Penrose generalized inverse module mapping as tool.

Theorem 1.4 *Let E be a Hilbert C^*-module over a C^*-algebra A, and let $T,\ S \in L(E)$. If $R(S)$ is closed in E then the following conditions are equivalent:*

(1) $R(T) \subset R(S)$;
(2) there exists a real number λ, such that $TT^* \leq \lambda^2 SS^*$;
(3) there exists a $Q \in L(E)$ such that $T = SQ$.

Proof (3) $\Rightarrow$ (1) is clear.

(1) $\Rightarrow$ (3): When x is in E, then $T(x) \in R(S)$. Since $R(S)$ is closed in E, $kerS$ is complemented in E. Therefore there exists uniquely $y \in ker(S)^\perp$ such that $T(x) = S(y)$. Define a A-linear mapping Q on E given by $Q : E \to E,\ x \to y$.

From the construction of Q, we have $T = SQ$. Since S has Moor-Penrose inverse $S^+ \in L(E)$ by Theorem 1.3, it follows that $S^+T = S^+SQ$. Since

$$ker(S)^{\perp} = R(S^*),$$

and

$$R(S^+S) = R(S^*),$$

from which implies that

$$S^+T(x) = S^+SQ(x)$$

$$= S^+S(y) = y = Q(x),$$

for all $x \in E$. Hence $Q = S^+T \in L(E)$.

(3) $\Rightarrow$ (2): From $T = SQ$ we have $TT^* = SQQ^*S^*$, which follows that $TT^* = SQQ^*S^*$. By the basic properties of positive cone in C^*-algebra, we have

$$QQ^* \leq ||Q||^2 I,$$

and

$$SQQ^*S^* \leq ||Q||^2 SS^*,$$

where I is the unit of $L(E)$, which implies from the above proof that $TT^* \leq \lambda^2 SS^*$, where $\lambda = ||Q||$.

(2) $\Rightarrow$ (3): Define a A-linear mapping $\widetilde{Q}$ on E given by

$$\widetilde{Q} : E \to E,\ \widetilde{Q}S^*(x) = T^*(x)\ (x \in E),\ \widetilde{Q}(x) = 0\ (x \in R(S^*)^{\perp}).$$

It follows that

$$||\widetilde{Q}(S^*x)||^2 = ||T^*x||^2$$

$$= || < T^*x,\ T^*x > || = || < TT^*x,\ x > ||$$

$$\leq \lambda^2 || < SS^*x,\ x > ||$$

$$= \lambda^2 ||S^*x||^2,$$

for all $x \in E$. Hence $\widetilde{Q}$ is well-defined and bounded. From the construction of $\widetilde{Q}$, we have $\widetilde{Q}S^* = T^*$. Since

$$S^*(S^*)^+S^* = S^*,\ and\ (S^+)^* = (S^*)^+,$$

and

$$ker(S^+)^* = ker S = R(S^*)^\perp.$$

It shows that $ker(S^*)^+ = R(S^*)^\perp$. Thus $\widetilde{Q} = \widetilde{Q}S^*(S^*)^+$. Therefore,

$$\begin{aligned}\widetilde{Q} &= \widetilde{Q}S^*(S^*)^+ \\ &= T^*(S^*)^+ = T^*(S^+)^* \in L(E).\end{aligned}$$

Finally, by performing $*$ operation on both sides of $\widetilde{Q}S^* = T^*$, it implies that $T = S\widetilde{Q}^*$. Hence $T = SQ$ if we write $Q = \widetilde{Q}^*$.

When the base space is reduced to the complex field $\mathbb{C}$, Theorem 1.4 becomes the Douglas factorization theorem:

Theorem 1.5 (see [4]) *Let H be a Hilbert space, and let $A,\ B$ be in $B(H)$. Then the following statements are equivalent:*

(1) $R(A) \subset R(B)$;
(2) $AA^* \leq \lambda^2 BB^*$ for some real number λ;
(3) $A = BC$ for some $C \in L(H)$.

From Theorem 1.4, we immediately get the following consequence.

Corollary 1.6 *If A is a C^*-algebra and elements a and b in A. Suppose that bA is a closed right ideal of A, then the following conditions are equivalent:*

(1) there exists an element $c \in A$ such that $a = bc$;
(2) there is $\lambda \geq 0$, such that $aa^* \leq \lambda^2 bb^*$.

Remark 1.21

(1) Corollary 1.3 had been considered by G.K. Pedersen [12] in the case of closed left ideals;
(2) In the proof process of Corollary 1.3, (1) $\Rightarrow$ (2) does not need the condition that bA is closed. However, in the proof of (2) $\Rightarrow$ (1) if bA being closed is removed, then the conclusion generally does not hold.

For example, Let A be $C[0,\ 1]$, which is an abelian C^*-algebra consisting of all continuous complex-valued functions on $[0,\ 1]$. Set

$$f(x) = |x - \frac{1}{2}|,\ g(x) = \begin{cases} f(x), & \text{if } x \in [0, \frac{1}{2}], \\ 2f(x), & \text{if } x \in [\frac{1}{2}, 1]. \end{cases}$$

Then $ff^* \leq gg^*$, but there is no $h \in C[0,\ 1]$ such that $f = gh$.

Remark 1.22 In [22], Xu Qingxiang and Sheng Lijun proved that a bounded adjointable C^* -linear operator between two Hilbert C^*-modules admits a bounded generalized inverse if and only if the operator has closed range.

Furthermore, inspired by the concept of Moore-Penrose generalized inverse module mapping, in [7], M. Frank and K. Sharifi discussed the generalized inverse of unbounded regular module operators.

At the end of this subsection, we are devoted to discussing Wold decomposition in the Hilbert C^*-module framework. It is well known that Wold decomposition is an important method for special orthogonal decomposition of Hilbert space by isometric operator. It is also the basis of unitary expansion of isometric operator. Moreover, the Nagy-Foias theory on unitary expansion of contraction operators is also based on it, so Wold decomposition plays an important role in operator theory.

In order to characterize Wold decomposition in Hilbert C^*-module framework, we still look for ideas in Wold decomposition of Hilbert space. Let H be a Hilbert space and T be in $B(H)$. If there exists a nontrivial closed subspace $H_0 \subset H$ such that $TH_0 \subset H_0$,, then we say that T has invariant subspace H_0. Moreover, if H_0 is the invariant subspace of T and T^* at the same time, then it is called the reduced subspace of T.

A closed proper subspace H_0 is called wandering for a given isometric operator $V \in B(H)$ if $V^n H_0 \perp V^m H_0$ when $n \neq m$ for all natural numbers n , m. Furthermore, if $H = \oplus_{n=0}^{\infty} V^n H_0$, then V is called a shift operator.

The above concepts can be transplanted to Hilbert C^*- modules in parallel and will not be repeated. In what follows, we review the meaning of infinite direct sums of Hilbert spaces and Hilbert C^*-modules, respectively:

If $\{H_n\}_{n=1}^{\infty}$ is a sequence of Hilbert spaces, then its direct sum is defined as

$$\oplus_{n=1}^{\infty} H_n = \left\{ (x_n)_{n=1}^{\infty} \in \prod_{n=1}^{\infty} H_n : \sum_{n=1}^{\infty} ||x_n||^2 < \infty \right\}.$$

Correspondingly, if $\{E_n\}_{n=1}^{\infty}$ is a sequence of Hilbert C^*-modules over a C^*-algebra A, then its direct sum is defined as

$$\oplus_{n=1}^{\infty} E_n = \left\{ (x_n)_{n=1}^{\infty} \in \prod_{n=1}^{\infty} E_n : \sum_{n=1}^{N} < x_n,\ x_n > \text{ is norm convergent in } A,\ N \in \mathbb{N} \right\},$$

which is a Hilbert C^*-module over C^*-algebra A when endowed with the following A-valued inner product:

$$< (x_n),\ (y_n) >= \sum_{n=1}^{\infty} < x_n,\ y_n >,$$

for all $(x_n)_{n=1}^{\infty},\ (y_n)_{n=1}^{\infty} \in \oplus_{n=1}^{\infty} E_n$.

In particular, if each $E_n = A,\ n = 1,\ 2,\ \cdots$, then $\{E_n\}_{n=1}^{\infty}$ becomes $l^2(A)$ mentioned above. In addition, if $\{F_n\}_{n=1}^{\infty}$ is a sequence of pairwise orthogonal closed submodules of Hilbert C^*-module E, then it is easy to prove that

$$\oplus_{n=1}^{\infty} F_n = \Big\{x = \sum_{n=1}^{\infty} x_n \text{ is norm convergent in } E, \text{ and } \sum_{n=1}^{\infty} < x_n,\ x_n > \text{ is}$$

$$\text{norm convergent in } A,\ x_n \in F_n,\ n \in \mathbb{N}\Big\}.$$

Definition 1.10 Let E be a Hilbert C^*-module and an isometric module operator V be in $L(E)$. If there exist two closed submodules E_0 and E_1 of E with the following properties:

(1) $E = E_0 \oplus E_1$;
(2) E_0 reduces V and $V|_{E_0}$ (this notation stands for the restriction of V on E_0) is unitary;
(3) $V|_{E_1}$ is a shift.

Then E is said to have a Wold type decomposition corresponding to V.

Recall that if V is an isometric operator V on Hilbert space H, then H has the following Wold decomposition:

$$H = \cap_{n=0}^{\infty} V^n H \oplus \oplus_{n=0}^{\infty} V^n (H \ominus VH).$$

D. Popovici [19] generalized the above conclusion to the framework of Hilbert C^*-modules and obtained the following Wold type decomposition theorem:

Theorem 1.6 *Let E be a Hilbert C^*-module and an isometric module operator V be in $L(E)$. Then E has a Wold type decomposition corresponding to V, if and only if the sequence $(< V^{*n}x,\ V^{*n} >)_{n=1}^{\infty}$ is norm convergent in A for all $x \in E$.*

Proof Suppose that for every $x \in E$, the sequence $(< V^{*n}x,\ V^{*n}x >)_{n=1}^{\infty}$ is norm convergent in A. In the following, we shall prove that $(V^n V^{*n} x)_{n=1}^{\infty}$ is a Cauchy sequence in E. Since for all $n,\ m \in \mathbb{N}$ and $n > m$, we have

$$\begin{aligned} < V^n V^{*n} x,\ V^n V^{*n} x > &= < V^{*n} V^n V^{*n} x,\ V^{*n} x > \\ &= < V^{*n} x,\ V^{*n} x >, \end{aligned}$$

and

$$< V^m V^{*m} x,\ V^m V^{*m} x > = < V^{*m} x,\ V^{*m} x > .$$

Also

$$< V^n V^{*n} x,\ V^m V^{*m} x >=< V^{*n} x,\ V^{*(n-m)} V^{*m} x >$$

$$=< V^{*n} x,\ V^{*n} x > .$$

Similarly

$$< V^m V^{*m} x,\ V^n V^{*n} x >=< V^{*n} x,\ V^{*n} x > .$$

Hence

$$||V^n V^{*n} x - V^m V^{*m} x||^2$$

$$= || < V^n V^{*n} x - V^m V^{*m} x,\ V^n V^{*n} x - V^m V^{*m} x > ||$$

$$= || < V^n V^{*n} x,\ V^n V^{*n} x > + < V^m V^{*m} x,\ V^m V^{*m} x > -$$

$$- < V^m V^{*m} x,\ V^n V^{*n} x > - < V^n V^{*n} x,\ V^m V^{*m} x > ||$$

$$= || < V^{*m} x,\ V^{*m} x > - < V^{*n} x,\ V^{*n} x > || \to 0,\ n,\ m \to \infty,$$

which shows that $(V^n V^{*n} x)_{n=1}^{\infty}$ is Cauchy sequence in E, so it is convergent with the limit $x_0 \in E$. It follows that $x_0 \in \cap_{n=0}^{\infty} V^n E$ because $V^n V^{*n} x \in \cap_{k=0}^{n} V^k E$ for all $n \in \mathbb{N}$. Let $L = ker V^*$. Since we have

$$< V^n l,\ V^m l' >=< V^{*m} V^n l,\ l' >$$

$$=< V^{n-m} l,\ l' >=< l,\ V^{*(n-m)} l' >= 0,$$

for all $n,\ m \in \mathbb{N}$ and $n > m$, and $l,\ l' \in L$. so that L is a wandering submodule for V.

We now prove that $E = \cap_{n=0}^{\infty} V^n E \bigoplus \oplus_{n=0}^{\infty} V^n L$. Since

$$E = ker V^* \oplus R(V)$$

$$= L \oplus VE = L \oplus VL \oplus V^2 E = \cdots$$

$$= L \oplus VL \oplus \cdots V^n L \oplus V^{n+1} E\ (n \in \mathbb{N}).$$

Therefore for any $x \in E$, there exist $\{l_k\}_{k=0}^{n} \subset L$ and $z_{n+1} \in E$, such that

$$x = \sum_{k=0}^{n} V^k l_k + V^{n+1} z_{n+1}.$$

Note that $VV^*V = V$, by simple calculation, we have then

$$(I_E - VV^*)x = \sum_{k=0}^{n}(I_E - VV^*)V^k l_k + (I_E - VV^*)V^{n+1}z_{n+1}$$

$$= (I_E - VV^*)l_0 = l_0.$$

Similarly

$$l_1 = (I_E - VV^*)V^*x;$$

$$l_2 = (I_E - VV^*)V^{*2}x;$$

$$\cdots\cdots\cdots\cdots\cdots$$

$$l_n = (I_E - VV^*)V^{*n}x;$$

$$z_{n+1} = V^{*(n+1)}x.$$

It follows that

$$\sum_{k=0}^{n} < l_k,\ l_k >= \sum_{k=0}^{n} < (I_E - VV^*)V^{*k}x,\ (I_E - VV^*)V^{*k}x >$$

$$= \sum_{k=0}^{n} < (I_E - VV^*)V^{*k}x, V^{*k}x >$$

$$= \sum_{k=0}^{n}(< V^{*k}x,\ V^{*k}x > - < V^{*(k+1)}x,\ V^{*(k+1)}x >)$$

$$=< x,\ x > - < V^{*(n+1)}x,\ V^{*(n+1)}x >,$$

which is norm convergent in A because $(< V^{*n}x,\ V^{*n}x >)_{n=1}^{\infty}$ is norm convergent in A. It shows that $\left(\sum_{k=0}^{n} < V^k l_k,\ V^k l_k >\right)_{n=1}^{\infty}$ is norm convergent in A since $\sum_{k=0}^{n} < V^k l_k,\ V^k l_k >= \sum_{k=0}^{n} < l_k,\ l_k >$. Since $\lim_{n\to+\infty} V^n V^{*n}x = x_0$, and

$$\sum_{k=0}^{n} V^k l_k = x - V^{n+1}z_{n+1} = x - V^{n+1}V^{*(n+1)},$$

from which it implies that

$$\sum_{n=0}^{\infty} V^n l_n = \lim_{n\to+\infty} \sum_{k=0}^{n} V^k l_k$$

$$= \lim_{n\to+\infty} (x - V^{n+1} V^{*(n+1)}) = x - x_0.$$

Above proof shows that $\sum_{n=0}^{\infty} V^n l_n \in \oplus_{n=0}^{\infty} V^n L$, and $x = x_0 + \sum_{n=0}^{\infty} V^n l_n$, where $x_0 \in \cap_{n=0}^{\infty} E$, $\sum_{n=0}^{\infty} V^n l_n \in \oplus_{n=0}^{\infty} V^n L$. Therefore,

$$E = \cap_{n=0}^{\infty} V^n E + \oplus_{n=0}^{\infty} V^n L$$

because x is an arbitrary element in E.

In what follows, For convenience, we write $\cap_{n=0}^{\infty} V^n E$ and $\oplus_{n=0}^{\infty} V^n L$ as E_0 and E_1, respectively. It is easy to check that $E_0 \perp E_1$. Therefore from the above proof that $E = E_0 \oplus E_1$. Moreover, E_0 and E_1 are reduced closed submodules for V. We next need to prove that $V|_{E_0}$ is unitary and $V|_{E_1}$ is shift. Since for any $x \in E_0 = \cap_{n=0}^{\infty} V^n E$, there exists $x' \in E$ such that $x = Vx'$. Hence,

$$(V|_{E_0})(V|_{E_0})^* x = (V|_{E_0})(V^*|_{E_0}) x$$

$$= (V|_{E_0})(V^* x) = VV^* x = VV^* V x' = V x' = x,$$

which shows that

$$(V|_{E_0})(V|_{E_0})^* = I_{E_0}.$$

Similarly

$$(V|_{E_0})^*(V|_{E_0}) = I_{E_0}.$$

Thus $V|_{E_0}$ is unitary. Since

$$E_1 = \oplus_{k=0}^{\infty} V^k L = \oplus_{k=0}^{\infty} (V|_{E_1})^k L,$$

and since L is a wandering closed submodule in E_1, $V|_{E_1}$ is a shift.

Conversely, suppose that $E = E_0 \oplus E_1$ is the Wold type decomposition corresponding to V, where $V_0 = V|_{E_0}$ is unitary, $V_1 = V|_{E_1}$ is a shift. It follows that there exists a wandering closed submodule L ($\subset E_1$) for V_1 such that $E_1 = \oplus_{n=0}^{\infty} V^n L$. It is easy to check that

$$V^* l = V_1^* l = 0,$$

for all $l \in L$. Hence for any $x \in E$, there exists a corresponding $x_0 \in E_0$ and $\{l_n\}_{n=0}^{\infty} \subset L$ satisfying $\sum_{n=1}^{\infty} < l_n, l_n >$ is norm convergent in A, such that

$$x = x_0 + \sum_{n=0}^{\infty} V^n l_n.$$

Since $V^n V^{*n} x = x$ for all $x \in E_0$ and all $n \in \mathbb{N}$, by simple calculation we have then

$$(I_E - VV^*)x = x_0 - VV^* x_0 + \sum_{n=0}^{\infty} (I_E - VV^*) V^n l_n = l_0;$$

$$(I_E - VV^*)V^* x = V^* x_0 - VV^{*2} x_0 + \sum_{n=0}^{\infty} (I_E - VV^*) V^n l_{n+1} = l_1;$$

$$\cdots\cdots\cdots\cdots\cdots$$

$$(I_E - VV^*)V^n x = l_n.$$

Therefore

$$< V^{*(n+1)},\ V^{*(n+1)} >=< x,\ x > - \sum_{k=0}^{n} < l_k,\ l_k > .$$

Thus $(< V^{*n} x,\ V^{*n} x >)_{n=0}^{\infty}$ is norm convergent in A because $\sum_{n=1}^{\infty} < l_n,\ l_n >$ converges in norm in A.

Remark 1.23

(1) In Theorem 1.6, since E_0 is a reduced closed submodule for V, then we have

$$\begin{aligned} < V|_{E_0}(x),\ y > &=< Vx,\ y >=< x,\ V^* y > \\ &=< x,\ V^*|_{E_0}(y) > \quad (x,\ y \in E_0), \end{aligned}$$

which shows that $(V|_{E_0})^* = V^*|_{E_0}$. Similarly, $(V|_{E_1})^* = V^*|_{E_1}$.

(2) It can be proved that if the Hilbert A-module E has a Wold type decomposition corresponding to an isometric module operator $V \in L(E)$, then the decomposition is unique, that is, there is only the following form:

$$E = \cap_{n=0}^{\infty} V^n E \oplus \oplus_{n=0}^{\infty} V^n L,$$

where $L = ker V^*$.

(3) Suppose that $E = E_0 \oplus E_1$ is the Wold type decomposition corresponding to an isometric module operator V, then it is easy to check that

$$E_0 = \{x \in E : < V^{*n}x,\ V^{*n}x >=< x,\ x >,\ (n \in \mathbb{N})\}$$

and

$$E_1 = \{x \in E :\ V^{*n}x \to 0,\ n \to \infty\}.$$

(4) If E is reduced to a Hilbert space, since the nonnegative sequence $(< V^{*n}x,\ V^{*n}x >)_{n=1}^{\infty}$ convergent obviously, Theorem 1.6 becomes the classical Wold decomposition theorem.

1.2.3 Topics on Module Operator Equations

Operator (matrix) equations have important applications in engineering problems, information theory, linear system theory, linear estimation theory, numerical analysis, and other topics. In the past two decades, a lot of achievements have been made in the study of module operator equations. Generalized inverse module mapping is one of the main tools to discuss the existence of solutions of module operator equations. The following is a brief introduction to the achievements in this regard; interested readers can refer to reference [1–3, 5, 20–23, 28, 29], etc.

Let E be a Hilbert C^*-module over a C^*-algebra A, and let $T,\ T',\ S,\ S'$ be all in $L(E)$. We first study the common solution to the module operator equations given by

$$TX = T',\ XS = S'.$$

Proposition 1.10 *Suppose that $\overline{R(T^*)}$ and $\overline{R(S)}$ are closed complemented submodules of E. Then*

$$TX = T,\ XS = S,\ \ (I)$$

have a common solution $X \in L(E)$, if and only if

$$R(T') \subset R(T),\ R(S'^*) \subset R(S^*),\ \text{and}\ TS' = T'S.$$

In this case, the general common solution is of the form

$$X = Y_1 + Y_2^* - PY_2^* + C,$$

where $Y_1,\ Y_2$ are the reduced solutions of $TX = T,\ S^*X = S'^*$, respectively. P is the projection of E onto $\overline{R(T^*)}$, and C is any operator in $L(E)$ with $R(C) \subset N(T),\ R(S) \subset N(C)$.

Proof We first prove the sufficiency. By the assumption, there exist orthogonal decompositions as follows:

$$E = \overline{R(T^*)} \oplus N(T),\ \ E = \overline{R(S)} \oplus N(S^*).$$

It follows from Theorem 1.4 that there exist operators $Y_1,\ Y_2 \in L(E)$ such that

$$TY_1 = T',\ \textit{and}\ R(Y_1) \subset N(T)^{\perp};$$

$$Y_2^*S = S',\ \textit{and}\ R(Y_2) \subset N(S^*)^{\perp}.$$

Indeed, Y_1 and Y_2 are the reduced solutions of equations $TX = T',\ \ S^*X = S'^*$, respectively. Let P be the projection of E onto $\overline{R(T^*)}$, then we have $T(I-P)S' = 0$, since $R(I-P) \subset N(T)$ from the orthogonal decompositions of E. Hence

$$TPS' = TS' = T'S = TY_1S.$$

From which it implies that $T(PS' - Y_1S) = 0$, namely, $R(PS' - Y_1S) \subset N(T)$. Note that $R(PS' - Y_1S) \subset N(T)^{\perp}$, so $PS' = Y_1S$. Furthermore, let $X = Y_1 + Y_2^* - PY_2^*$. Simple calculation shows that X is the solution of the above equation group (I).

Next, to show the necessity. For this, by Theorem 1.4 again, we have $R(T') \subset R(T),\ \ R(S'^*) \subset R(S^*)$, and $TS' = TXS = T'S$. So we are done.

Remark 1.24 The existence of hermitian solutions and positive solutions of the above equations group was considered in [29].

In the following, we shall consider the solvability of the operator equation $TXS^* - SX^*T^* = K$ for adjointable operators on Hilbert C*-module.

Definition 1.11 Let E be a Hilbert C^*-module over a C^*-algebra A, and let T be in $L(E)$. If A-linear operator $T^- : D(T^-) \to E$ satisfies that

$$TT^-T = T,$$

then T^- is said to be an inner inverse of T.

Remark 1.25 Even in the case of Hilbert space, the inner inverse of a bounded linear operator T is not necessarily bounded if it exists, and it is generally not unique. It is easy to prove that the inner inverse of T is bounded if and only if $R(T)$ is closed. This conclusion is also valid in the framework of Hilbert C^*-modules.

Proposition 1.11 *Suppose that T and S are in $L(E)$ with closed ranges, and $R(S) \subset R(T)$. Let K be in $L(E)$. Then the following equation*

$$TXS^* - SX^*T^* = K,$$

has a solution in $L(E)$ if and only if

$$K^* = -K, \text{ and } (TT^+ + S^-T(S^-T)^+)KSS^+ - ((TT^+ + S^-T(S^-T)^+)KSS^+)^* = 2K.$$

The proof of Proposition 1.11 makes comprehensive use of the definitions and properties of generalized inverse and inner inverse for calculation. The specific calculation process can be found in reference [23]. In [23], they also discussed the solution to the operator equation $TXS = K$ and obtain some necessary and sufficient conditions for the existence of a real positive solution, of a solution X with $S^*(X^* + X)S \geq 0$, and of a solution X with $S^*XS \geq 0$. Furthermore in the special case that $R(S) \subset \overline{R(T^*)}$, a necessary and sufficient condition for the existence of a positive solution to the equation $TXS = K$ is obtained.

In [5], F.O. Farid, M.S. Moslehian, Q.W. Wang and Z.C. Wu synthesized the previous results and investigated the Hermitian solution to the following broader and more complex system of adjointable operator equations:

$$A_1X_1 = C_1,\ X_1B_1 = D_1,\ A_2X_2 = C_2,\ X_2B_2 = D_2, A_3X_1A_3^* + A_4X_2A_4^* = C_3\ (II).$$

Moreover, Chang-Zhou Dong, Qing-Wen Wang, and Yu-Ping Zhang in [2] discussed the existence of the common positive solution to the above equations (II), obtained necessary and sufficient conditions for the existence of the common positive solution to the above equations (II) for adjointable operators between Hilbert C^*-modules, and present an expression of the positive solution to the system when the solvability conditions are satisfied.

For a given Hilbert C^*-module E, let $A_1,\ A_2,\ A_1,\ A_4,\ B_2,\ B_2,\ C_1,\ C_2,\ C_3,\ D_1,\ D_2$ be all in $L(E)$. For the sake of brevity discussed below, several notations are specified here. Set

$$T_1 = \begin{pmatrix} A_1 \\ B_1^* \end{pmatrix},\ S_1 = \begin{pmatrix} C_1 \\ D_1^* \end{pmatrix},\ K_1 = \begin{pmatrix} C_1A_1^* & C_1B_1 \\ (A_1D_1)^* & D_1^*B_1 \end{pmatrix}$$

$$T_2 = \begin{pmatrix} A_2 \\ B_2^* \end{pmatrix},\ S_1 = \begin{pmatrix} C_2 \\ D_2^* \end{pmatrix},\ K_1 = \begin{pmatrix} C_2A_2^* & C_2B_2 \\ (A_2D_2)^* & D_2^*B_2 \end{pmatrix}$$

$$G = C_3 - A_3S_1^*K_1^-S_1A_3^* - A_4S_2^*K_2^-S_2A_4^*,\ M_1 = A_3(I - E_1^-E_1),$$
$$M_2 = A_4(E_2^-E_2),$$

$$M_3 = (I - M_1 M_1^+) M_2,\ M_4 = (I - M_1 M_1^+) G,\ M = M_3^+ M_3,\ N = M_2^*(M_2^*)^+,$$

$$P = M_3^+ M_4 (M_2^*)^+,\ Q = M + N, R_1 = Q^+ N,\ R_2 = M Q^+,\ J = N Q^+ P Q^+ M.$$

Proposition 1.12 *The meanings of $A_1,\ A_2,\ A_3,\ A_4,\ B_2,\ B_3,\ C_1,\ C_2,\ C_3,\ D_1,\ D_2$ are as above, and suppose that $T_1,\ T_2,\ K_1,\ K_2,\ M_1,\ M_2,\ M_3,\ M,\ N,\ R_1,\ R_2,\ J$ have closed ranges, respectively. Then the following statements are equivalent:*

(1) Equations (II) have a pair of positive solutions $X_1,\ X_2 \in L(E)$;
(2) $C_3 \geq 0,\ K_1 \geq 0,\ K_2 \geq 0,\ R(S_1) \subset R(K_1), R(S_2) \subset R(K_2)$, and $(I - M_1 M_1^+) G (I - M_2 M_2^+) = 0,\ (I - M_3 M_3^+)(I - M_1 M_1^+) G = 0,$ $M M^+ P N^+ N = P,\ J \geq 0,\ R(M Q^+ P^*) \subset R(J),\ R(N Q^+ P) \subset R(J).$

The proof of Proposition 1.12 can be found in [20]. Moreover, in [20], a specific expression of the solution to above equation system (II) was given.

Remark 1.26 Recently, by using two well-known results of Douglas and Sebestyén, Dragana Cvetković-Ilić, Qing-Wen Wang, and Qingxiang Xu in [3] presented some equivalents of the existence of a positive solution of the operator equation $AXB = C$ without any additional range assumptions.

1.3 Tensor Products of Hilbert C^*-Modules

In order to characterize the tensor products of Hilbert C^*-modules, we briefly review algebraic tensor products of linear spaces (containing Hilbert spaces and C^*-algebras) and related concepts. In fact, algebraic tensor product has appeared in the previous example.

Suppose that V and W are finite dimensional linear spaces and V^* and W^* are their algebraic, respectively. Let $v \in V,\ w \in W$, define a bilinear function from $V^* \times W^*$ to $\mathbb{C}$ given by:

$$v \otimes w :\ V^* \times W^* \to \mathbb{C}, v \otimes w(f,\ g) = f(v) g(w),$$

for all $f \in V^*,\ g \in W^*$. Then the tensor product of V and W is the vector space generated by the linear combination of elements of the form $v \otimes w(v \in V,\ w \in W)$, which is denoted by $V \otimes W$. Let $(v_i)_{i=1}^n$ and $(w_j)_{j=1}^m$ be bases of V and W, respectively. Then $V \otimes W$ is generated linearly by $\{v_i \otimes w_j :\ i = 1,\ \cdots,\ n;\ j = 1,\ \cdots,\ m\}$ whose called a base of $V \otimes W$, from which it shows that the dimension of $V \otimes W$ is nm. Equivalently, let

$$\otimes : M_{m,\ 1} \times M_{n,\ 1} \to M_{m,\ n},\ x \otimes y = x y^T,$$

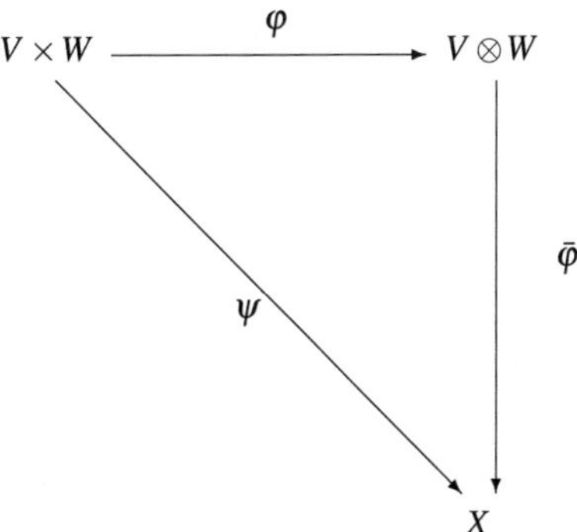

Fig. 1.2 Commutative graph

for all $x \in M_{m,\,1},\ \ y \in M_{n,\,1}$, where $M_{m,\,1}$ and $M_{n,\,1}$ stand for the set of all $m \times 1$ and $n \times 1$ matrices, respectively. It follows that $M_{m,\,1} \otimes M_{n,\,1} = M_{m,\,n}$.

In general, let $(e_i)_{i\in I}$ and $(f_j)_{j\in J}$ be bases of (infinite dimension) linear spaces X and Y, respectively. Then the vector space generated linearly by $(e_i \otimes f_j : i \in I, j \in J)$ as base is called the tensor product of V and W, which is denoted by $V \otimes W$.

The above definition of tensor product is equivalent to that the bilinear mapping $\varphi : V \times W \to V \otimes W$ has universal property (here $V \otimes W$ represents only a linear space) as follows:

For any given linear space X and any bilinear mapping $\psi :\ \ V \times W \to X$, there is only linear mapping $\overline{\varphi} : V \otimes W \to X$ which makes the following figure commutative (Fig. 1.2):

If X and Y are normed linear spaces, we denote by $X \otimes_{alg} Y$ their algebraic tensor product. There are in general many possible norms on $X \otimes_{alg} Y$. But for Hilbert spaces H and K, there is a unique norm on $H \otimes_{alg} K$ which is induced by the following inner product:

$$< x \otimes y,\ x' \otimes y' >=< x,\ x' >< y,\ y' >,$$

for all $x,\ \ x' \in H$ and $y,\ \ y' \in K$. Therefore the completion $H \otimes K$ of $H \otimes_{alg} K$ with respect to the above norm is the tensor product of H and K. $H \otimes K$ is also a Hilbert space and satisfies that $||x \otimes y|| = ||x||||y||$ for all $x \in H$ and $y \in K$. We call the norm satisfying this equation a cross norm.

If A and B are $*$ algebras, denote by $A \otimes_{alg} B$ their algebraic tensor product as linear spaces. There are a unique multiplication and a unique involution $*$ on $A \otimes_{alg} B$ given by $(a \otimes b)(a' \otimes b') = aa' \otimes bb'$ for all $a,\ a'$ in A and $b,\ b'$ in B, and $(a \otimes b)^* = a^* \otimes b^*$ for all a in A and b in B. Therefore $A \otimes_{alg} B$ becomes a $*$ algebra with respect to above multiplication and involution on it.

Furthermore, if A and B are C^*-algebras, then we can define different kinds of cross norms on $A \otimes_{alg} B$ such that its completion is a C^*-algebra. In this case, such cross norms are called C^*-norms on $A \otimes_{alg} B$. Among them there is a minimal and a maximal C^*-norm defined as follows:

$$||x||_{min} = \sup ||(\pi_1 \otimes \pi_2)(x)|| \ (x \in A \otimes_{alg} B),$$

where π_1 and π_2 run over all representations of A and B, respectively. The completion $A \otimes_{min} B$ is called the injective or spatial C^*-tensor product of A and B. The maximal C^*-norm is given by

$$||x||_{max} = \sup ||\pi(x)|| \ (x \in A \otimes_{alg} B),$$

where π runs throughout all representations of $A \otimes_{alg} B$. The completion $A \otimes_{max} B$ is called the projective C^*-tensor product of A and B.

In this book, we only use the spatial C^*-norm, so we rewrite $A \otimes_{min} B$ as $A \otimes B$ for convenience.

1.3.1 Exterior Tensor Product of Hilbert C*-Modules

Let A and B be C^*-algebras. Suppose that E is a Hilbert A-module and F is a Hilbert B-module. Then $E \otimes_{alg} F$ is a right $A \otimes_{alg} B$-module with the module action given by

$$(e \otimes f)(a \otimes b) = ea \otimes fb$$

for all $e \in E, \ f \in F$ and $a \in A, \ b \in B$. Next, define a $A \otimes_{alg} B$-valued sesquilinear form $<, >$ on $E \otimes_{alg} F$ which is the linear extension of the following form:

$$< e_1 \otimes f_1, \ e_2 \otimes f_2 >=< e_1, \ e_2 > \otimes < f_1, \ f_2 >,$$

for all $e_1 \otimes f_1$ and $e_2 \otimes f_2$ in $E \otimes_{alg} F$. We now prove that it is nonnegative definite. Let

$$x = \sum_{i=1}^{k} e_i \otimes f_i \ (e_i \in E, \ f_i \in F, \ i = 1, \ \cdots, \ k).$$

Since for all $b_1, \ \cdots, \ b_k \in A$, we have then

$$\sum_{i,j} b_i^* < e_i, \ e_j > b_j =< \sum_i e_i b_i, \ \sum_i e_i b_i >\geq 0.$$

Hence by M. Takeski [27, IV.Lemma 3.2] (referring to the appendix of this book also), matrix $(< e_i,\ e_j >)_{i,j}$ is positive in $M_k(A)$, so that there exists matrix $(a_{ij}) \in M_k(A)$ such that

$$< e_i,\ e_j >= \sum_{n=1}^{k} a_{ni}^* a_{nj}\ (i,\ j = 1,\ 2,\ \cdots,\ k).$$

Similarly, there exists matrix $(b_{ij}) \in M_k(B)$ such that

$$< f_i,\ f_j >= \sum_{n=1}^{k} b_{ni}^* b_{nj}\ (i,\ j = 1,\ 2,\ \cdots,\ k),$$

from which it implies that

$$< x,\ x >= \sum_{i,\ j} < e_i,\ e_j > \otimes < f_i,\ f_j >$$

$$= \sum_{i,\ j} \sum_{m,\ n} a_{ni}^* a_{nj} \otimes b_{mi}^* b_{mj}$$

$$= \sum_{m,\ n} \left(\sum_i a_{ni} \otimes b_{mi} \right)^* \left(\sum_i a_{ni} \otimes b_{mi} \right) \geq 0.$$

Therefore, the bilinear form $<,\ >$ on $E \otimes_{alg} F$ is nonnegative defined. Let $N = \{x \in E \otimes_{alg} F :< x,\ x >= 0\}$, then $E \otimes_{alg} F/N$ is a right $A \otimes_{alg} B$-module with $A \otimes_{alg} B$-valued inner product by the following way:

$$< e_1 \otimes f_1 + N,\ e_2 \otimes f_2 + N >=< e_1 \otimes f_1,\ e_2 \otimes f_2 >,$$

for all $e_1 \otimes f_1 + N$ and $e_2 \otimes f_2$ in $E \otimes_{alg} F/N$. Furthermore, since

$$< (e \otimes f + N)(a \otimes b),\ (e \otimes f + N)(a \otimes b) >$$

$$=< ea \otimes fb + N,\ ea \otimes fb + N >$$

$$=< ea \otimes fb,\ ea \otimes fb >$$

$$=< (e \otimes f)(a \otimes b),\ (e \otimes f)(a \otimes b) >,$$

for all $e \otimes f \in E \otimes_{alg} F$ and $a \otimes b \in A \otimes_{alg} B$. It follows that

$$||(e \otimes f + N)(a \otimes b)|| = ||(e \otimes f)(a \otimes b)||$$
$$\leq ||e \otimes f|| \cdot ||a \otimes b||.$$

Hence the completion of $E \otimes_{alg} F/N$ is a Hilbert $A \otimes_{alg} B$-module, which results from the above inequality that $E \otimes F$ can be extended by continuity to Hilbert $A \otimes B$-module, denoted by $E \otimes F$, which is called the exterior tensor product of E and F.

Remark 1.27 It can be proved by using the Kasprove' stability theorem (see Sect. 2.1 in Chap. 2) that the above mentioned $N = \{0\}$. Thus the semi-inner product defined above is exactly the inner product. Refer to E.C. Lance [13], P_{62} for details.

Example 1.8 Suppose that A is a C^*-algebra and H is a separable Hilbert space. Then the algebraic tensor product $H \otimes_{alg} A$ is a pre-Hilbert A-module when endowed with the following A-valued inner product:

$$< \xi \otimes a, \ \eta \otimes b > = < \xi, \ \eta > a^* b,$$

for all $\xi, \ \eta \in H$ and $a, \ b \in A$. Its completion is a Hilbert A-module, which is denoted by $H \otimes A$.

We now show that $H \otimes A$ is unitary equivalent to $l^2(A)$ defined previously. Let $\{e_i\}_{i=1}^{\infty}$ be standard orthogonal base of H. For any fixed elements $\xi = \sum_{i=1}^{\infty} \lambda_i e_i \in H$ and $a \in A$, define a A- linear mapping U from $H \otimes A$ to $l^2(A)$ given on simple tensors by

$$U(\xi \otimes a) = (\lambda_i a)_{i=1}^{\infty}.$$

It is easy to check that U is unitary, namely, $H \otimes A \cong l^2(A)$.

Remark 1.28 $l^2(A)$ can be used as the representative of $H \otimes A$. It plays a key role in Kasprove' stability theory, Fredholm module operator theory and quantum Markov module operator semigroups. It is pointed out that we choose to use $l^2(A)$ or $H \otimes A$ according to the needs of the application in the later.

Suppose that A and B are C^*-algebras. From Example 1.5 we see that $K(A) \cong A$ and $K(B) \cong B$. Hence

$$K(A \otimes B) \cong A \otimes B \cong K(A) \otimes K(B).$$

As for Hilbert spaces H_1 and H_2, there is a similar conclusion as follows:

$$K(H_1 \otimes H_2) \cong K(H_1) \otimes K(H_2).$$

The above conclusion can be extended to the case of Hilbert C^*-module, that is,

Proposition 1.13 *Suppose that E and F are Hilbert A-modules. Then*

$$K(E) \otimes K(F) \cong K(E \otimes F).$$

Proof Since $\theta_{u,\, v} \otimes \theta_{x,\, y} \in K(E) \otimes K(F)$, and $\theta_{u\otimes x,\, v\otimes y} \in K(E \otimes F)$ for all $u,\ v \in E$ and $x,\ y \in F$. Then we can define a A-linear mapping U from $K(E) \otimes K(F)$ to $K(E \otimes F)$ given on simple tensors by

$$U : \theta_{u,\, v} \otimes \theta_{x,y} \to \theta_{u\otimes x,\, v\otimes y}.$$

On the one hand, by direct calculation we have

$$< U(\theta_{u,\, v} \otimes \theta_{x,\, y}), U(\theta_{u',\, v'} \otimes \theta_{x',\, y'}) >$$

$$=< \theta_{u\otimes x,\, v\otimes y}, \theta_{u'\otimes x',\, v'\otimes y'} >$$

$$= (\theta_{u\otimes x,\, v\otimes y})^* (\theta_{u'\otimes x',\, v'\otimes y'})$$

$$= \theta_{v\otimes y,\, u\otimes x} \theta_{u'\otimes x',\, v'\otimes y'}$$

$$= \theta_{v\otimes y,\, \theta^*_{u'\otimes x',\, v'\otimes y'}(u\otimes x)}$$

$$= \theta_{v\otimes y,\, \theta_{v'\otimes y',\, u'\otimes x'}(u\otimes x)}$$

$$= \theta_{v\otimes y,\, v'\otimes y' <u'\otimes x',\, u\otimes x>}.$$

On the other hand, since

$$< \theta_{u,\, v} \otimes \theta_{x,\, y}, \theta_{u',\, v'} \otimes \theta_{x',\, y'} >$$

$$= (\theta_{u,\, v} \otimes \theta_{x,\, y})^* (\theta_{u',\, v'} \otimes \theta_{x',\, y'})$$

$$= (\theta^*_{u,\, v} \otimes \theta^*_{x,\, y})(\theta_{u',\, v'} \otimes \theta_{x',\, y'})$$

$$= (\theta_{v,\, u} \otimes \theta_{y,\, x})(\theta_{u',\, v'} \otimes \theta_{x',\, y'})$$

$$= \theta_{v,\, u} \theta_{u',\, v'} \otimes \theta_{y,\, x} \theta_{x',\, y'}$$

$$= \theta_{v,\, \theta_{v',\, u'}(u)} \otimes \theta_{y,\, \theta_{y',\, x'}(x)}$$

$$= \theta_{v,\, v' <u',\, u>} \otimes \theta_{y,\, y' <x',\, x>}$$

$$= \theta_{v\otimes y,\, v'\otimes y' <u'\otimes x',\, u\otimes x>},$$

from which it implies that

$$< U(\theta_{u,\,v} \otimes \theta_{x,\,y}), U(\theta_{u',\,v'} \otimes \theta_{x',\,y'}) >$$

$$=< \theta_{u,\,v} \otimes \theta_{x,\,y}, \theta_{u',\,v'} \otimes \theta_{x',\,y'} > .$$

Hence U is isometric. Since $span\{\theta_{u,\,v} \otimes \theta_{x,\,y} \,:\, u,\; v \in E,\; x,\; y \in F\}$ is dense in $K(E) \otimes K(F)$, U can be extended by continuity to an isometric mapping from $K(E) \otimes K(F)$ to $K(E \otimes F)$, and it is easy to check that U is $*$ operation preserving mapping. Define a module mapping $V : K(E \otimes F) \to K(E) \otimes K(F)$ given on the simple tensors by

$$\theta_{u \otimes x,\, v \otimes y} \to \theta_{u,\,v} \otimes \theta_{x,y}.$$

It is easy to prove that V is isometric. Therefore by the constructions of U and V, we have

$$UV = I_{K(E \otimes F)};\;\; VU = I_{K(E) \otimes K(F)}.$$

Hence U is a $*$ preserving isomorphic mapping from $K(E) \otimes K(F)$ onto $K(E \otimes F)$, namely,

$$K(E) \otimes K(F) \cong K(E \otimes F).$$

Example 1.9 Given Hilbert C^*-module $H \otimes A$. Then we have

$$K(H \otimes A) \cong K(H) \otimes A.$$

In fact, the above conclusion can be obtained immediately by Proposition 1.13 combined with Example 1.5.

Remark 1.29 Let $j : K(E \otimes F) \to M(K(E \otimes F)) = L(E \otimes F)$ be the embedding mapping. By Proposition 1.13, there exists an isometric $*$ isomorphism $U : K(E) \otimes K(F) \to K(E \otimes F)$, from which it shows that the composition mapping $\Phi = j \circ U : K(E) \otimes K(F) \to L(E \otimes F)$, and is also isometric $*$ isomorphism. Since $K(E)$ and $K(F)$ are closed essential ideals in $L(E)$ and $L(F)$, respectively, it is easy to check that $K(E) \otimes K(F)$ is a closed essential ideal in $L(E) \otimes L(F)$. Hence by Theorem 3.1.8 in G.J. Murphy [17], there exists uniquely $*$ isomorphism $\widetilde{\Phi}$: $L(E) \otimes L(F) \to L(E \otimes F)$ extending the above Φ.

However, the above isomorphism $\widetilde{\Phi}$ is generally not surjective, even if E and F are Hilbert spaces. But if at least one of them is a finite dimensional space, then we can obtain the following conclusion:

Proposition 1.14 *Suppose that H and K are Hilbert spaces. Then $B(H) \otimes B(K) \cong B(H \otimes K)$ if and only if at least one of H and K is a finite dimensional space.*

Proof If $K = \mathbb{C}^n$, where n is a natural number, then

$$H \otimes K = H \otimes C^n \cong H^n,$$

where $H^n = H \oplus H \oplus \cdots \oplus H$. Hence,

$$B(H \otimes K) \cong B(H^n) = M_n(B(H)).$$

Furthermore, since $B(K) = B(C^n) = M_n(C)$, we have then

$$B(H) \otimes B(K) = B(H) \otimes M_n(C) \cong M_n(B(H)).$$

So that

$$B(H) \otimes B(K) \cong B(H \otimes K).$$

Conversely, assume that H and K are infinite dimensional Hilbert spaces. We denote by $B(H)\overline{\otimes}B(K)$ the von Neumann algebra tensor product of $B(H)$ and $B(K)$. From Theorem 4.3.4 in [15], we have

$$B(H)\overline{\otimes}B(K) = B(H \otimes K).$$

Note that $B(H) \otimes B(K)$ is dense in $B(H)\overline{\otimes}B(K)$ with respect to the weak $*$-topology by Theorem 4.3.3 in [15], which follows that $B(H) \otimes B(K)$ is a proper subset of $B(H \otimes K)$.

1.3.2 Interior Tensor Product of Hilbert C-Modules*

The interior tensor product of Hilbert C^*-modules is constructed by a $*$ homomorphism or a completely positive mapping. Next we shall give the concrete construction method:

Let E and F be Hilbert C^*-modules over a C^*-algebra A, and let $\phi : A \to L(F)$ be a given $*$ homomorphism. Then algebraic tensor product $E \otimes_{alg} F$ is a right A-module, which the module action is given by

$$(x \otimes y)a = x \otimes ya,$$

for all $x \in E$, $y \in F$ and $a \in A$. Then we define a A-value mapping $<,\ >$: $E \otimes_{alg} F \times E \otimes_{alg} F \to A$ given by

$$< x_1 \otimes x_2,\ y_1 \otimes y_2 >=< x_2,\ \phi(< x_1,\ y_1 >)y_2 >,$$

for all $x_1, y_1 \in E$ and $x_2, y_2 \in F$. Since $\phi(< x_1, y_1 >) \in L(F)$, then we have

$$< x_2, \phi(< x_1, y_1 >)y_2 >$$

$$=< \phi^*(< x_1, y_1 >)x_2, y_2 >$$

$$=< \phi(< x_1, y_1 >^*)x_2, y_2 >$$

$$=< \phi(< y_1, x_1 >)x_2, y_2 >,$$

from which it shows that

$$< x_1 \otimes x_2, y_1 \otimes y_2 >^* =< y_1 \otimes y_2, x_1 \otimes x_2 > .$$

Hence the above defined mapping $<, >$ is a A-value sesquilinear form. Next, to show that $<, >$ is nonnegative. Let $z = \sum_{i=1}^{n} x_i \otimes y_i \in E \otimes_{alg} F$. Then we have

$$< z, z >= \sum_{i,j} < y_i, \phi(< x_i, x_j >)y_j > .$$

By IV. Lemma 3.2 in [27] (see also the appendix of this book) matrix $(< x_i, x_j >)_{ij}$ is a positive element in $M_n(A)$. Since ϕ is $*$ homomorphism, ϕ is completely positive map. Hence $\phi^{(n)}((< x_i, x_j >))$ is a positive element in $L(F^n)$. Write $y = (y_1, \cdots, y_n) \in F^n$,, then we have

$$< z, z >=< y, \phi^{(n)}((< x_i, x_j >))y >$$

$$=< [\phi^{(n)}((< x_i, x_j >))]^{1/2}, [\phi^{(n)}((< x_i, x_j >))]^{1/2} >\geq 0.$$

Let $N = \{z \in E \otimes_{alg} F :< z, z >= 0\}$, then N is a right A-module, and the quotient space $E \otimes_{alg} F/N$ is a pre-Hilbert A-module when endowed with the following A-valued inner product:

$$< x_1 \otimes x_2 + N, y_1 \otimes y_2 + N >=< x_2, \phi(< x_1, y_1 >)y_2 >,$$

for all $x_1, y_1 \in E$ and $x_2, y_2 \in F$. Thus, the completion of $E \otimes_{alg} F/N$ with the norm induced by the above A-valued inner product is called interior tensor product of E and F corresponding to ϕ.

Remark 1.30

(1) Denote by N_0 the subspace generated by elements of the form $(xa \otimes y - x \otimes \phi(a)y$ $(x \in E,\ y \in F,\ a \in A)$. From the proof of Proposition 4.5 in [13], the above N is equal to N_0, which shown that the interior tensor product is balanced with respect to C^*-algebra A, this means that

$$< xa \otimes y,\ z >=< x \otimes \phi(a)y,\ z >,$$

for all $x \in E,\ y \in F$ and $a \in A,\ z \in E \otimes_{alg} F$.

(2) If E is a Hilbert C^*-module over a C^*-algebra $A,\ \ F$ is a Hilbert C^*-module over a C^*-algebra B, and we are given a $*$ homomorphism $A \to L_B(F)$. Then the interior tensor product can be defined similarly, where $L_B(F)$ is intended to emphasize that F is a right B-module.

Example 1.10 Suppose that A is a C^*-algebra and H is a Hilbert space which is regarded as a Hilbert module over the complex field $\mathbb{C}$. If given a mapping φ from $\mathbb{C}$ to $L(A)$ in the following way:

$$\varphi : \mathbb{C} \to L(A) = M(A),\ \varphi(\lambda)a = \lambda a\ (\lambda \in \mathbb{C},\ a \in A).$$

Then

$$H \otimes_\varphi A \cong H \otimes A.$$

In fact, let $U : x \otimes_\varphi a \to x \otimes a\ (x \in H,\ a \in A)$. Since

$$< U(x \otimes_\varphi a),\ U(y \otimes_\varphi b) >$$

$$=< x \otimes a,\ y \otimes b >$$

$$=< x,\ y > a^*b,$$

Also

$$< x \otimes_\varphi a,\ y \otimes_\varphi b >$$

$$=< a,\ \varphi(< x,\ y >)b >$$

$$= \varphi(< x,\ y >) < a,\ b >$$

$$=< x,\ y > a^*b.$$

It follows that

$$< U(x \otimes_\varphi a),\ U(y \otimes_\varphi b) >=< x \otimes_\varphi a,\ y \otimes_\varphi b >,$$

this indicates that the mapping U preserves the inner product operation. Note that the range $R(U)$ is dense in $H \otimes A$ and the set $span\{x \otimes_{\varphi} a : x \in H,\ a \in A\}$ is dense in $H \otimes_{\varphi} A$, so U can be extended to an isometric surjection from $H \otimes_{\varphi} A$ onto $H \otimes A$, namely, U is unitary, in which denoted by $x \otimes_{\varphi} a = x \otimes a + N \in H \otimes_{alg} A/N$.

Example 1.11 Suppose that A and B are C^*-algebras and H is a Hilbert space. If F is a Hilbert B-module and there exists a $*$ homomorphism $\phi : A \to M(B)$ such that $\overline{\phi(A)B} = B$, then $(H \otimes A) \otimes_{\phi} B \cong H \otimes B$, where the bar $(-)$ in the above indicates norm closure in B.

Indeed, let $U : (\xi \otimes a) \otimes_{\phi} b \to \xi \otimes \phi(a)b\ (\xi \in H,\ a \in A,\ b \in B)$. When $\eta \in H,\ a' \in A,\ b' \in B$, since

$$
\begin{aligned}
&< U((\xi \otimes a) \otimes_{\phi} b),\ U((\eta \otimes a') \otimes_{\phi} b') > \\
&=< \xi \otimes \phi(a)b,\ \eta \otimes \phi(a')b' > \\
&=< \xi,\ \eta >< \phi(a)b,\ \phi(a')b' > \\
&=< \xi,\ \eta >< b,\ \phi(a^*a')b' > \\
&=< \xi,\ \eta >< b,\ \phi(< a,\ a' >)b' > .
\end{aligned}
$$

Also

$$
\begin{aligned}
&< (\xi \otimes a) \otimes_{\phi} b,\ (\eta \otimes a') \otimes_{\phi} b' > \\
&=< b,\ \phi(< \xi \otimes a,\ \eta \otimes a' >)b' > \\
&=< b,\ \phi(< \xi,\ \eta >< a,\ a' >)b' > \\
&=< \xi,\ \eta >< b,\ \phi(< a,\ a' >)b' > .
\end{aligned}
$$

It follows that

$$
\begin{aligned}
&< U((\xi \otimes a) \otimes_{\phi} b),\ U((\eta \otimes a') \otimes_{\phi} b') > \\
&=< (\xi \otimes a) \otimes_{\phi} b,\ (\eta \otimes a') \otimes_{\phi} b' >,
\end{aligned}
$$

this shows that U keeps B-valued inner product operation. Since sums of elements of the form $(\xi \otimes a) \otimes_{\phi} b$ are dense in $(H \otimes A) \otimes_{\phi} B$, the range $R(U)$ of U is dense also. Thus U can be extended to an isometric surjection from $(H \otimes A) \otimes_{\phi} B$ to $H \otimes B$, so U is a unitary module operator.

1.4 The KSGNS Construction

1.4.1 GNS Construction and Stinespring Representation Theorem

We first review briefly the GNS construction in C^*-algebra theory. It is well known that GNS construction is the representation theorem of positive linear functional on C^*-algebra C^*-algebra, which reveals that every C^*-algebra can be regarded as a C^*-subalgebra C^*-algebra of $B(H)$ for some Hilbert space H.

Let A be a C^*-algebra and f be a given positive linear functional on A. Put $N_f = \{a \in A : f(a^*a) = 0\}$, by Cauchy-Schwarz inequality, it is easy to check that N_f is a closed left ideal of A. Hence the quotient space A/N_f is a pre-Hilbert space when endowed with the following inner product

$$<,>: A/N_f \times A/N_f \to \mathbb{C}, < a + N_f,\ b + N_f >= f(a^*b).$$

The completion of A/N_f is a Hilbert space, which is denoted by H_f. Let $B(A/N_f)$ denotes the set of all bounded linear operators on the inner product space A/N_f. Since

$$||\pi_f(a)(b + N_f)||^2 = ||ab + N_f||^2$$

$$=< ab + N_f,\ ab + N_f >= f(b^*a^*ab)$$

$$\leq ||a||^2 f(b^*b) = ||a||^2||b + N_f||,$$

which shows that $||\pi_f(a)|| \leq ||a||$. Note that A/N_f is dense in H_f, so $\pi_f(a)$ can be uniquely extended to a bounded linear operator on H_f, still write as $\pi_f(a)$. It is straightforward to check that $\pi_f : A \to B(H_f),\ a \to \pi_f(a)$ is a $*$ homomorphism.

When A has unit 1, write $\xi_f = 1 + N_f$. Observe that $\pi_f(a)\xi_f = \pi_f(a)(1 + N_f) = a + N_f\ (a \in A)$,; hence $\pi_f(A)\xi_f$ is dense in H_f and ξ_f is a cyclic vector for π_f, from which it shows that

$$f(a) =< 1 + N_f,\ a + N_f >=< \xi_f,\ \pi_f(a)\xi_f > \quad (a \in A).$$

Put $a = 1$ in the above equation, we have

$$1 = f(1) =< \xi_f,\ \pi_f(1)\xi_f >$$

$$=< \xi_f,\ \xi_f >= ||\xi_f||^2,$$

which indicates that the cyclic vector ξ_f is a unit vector.

When A has no unit, let $\{u_\lambda\}$ be a an approximate unit of A, so $\{\pi_f(u_\lambda)\}$ is a Cauchy net because $||\pi_f(a)|| \leq ||a||$ for all $a \in A$, which follows that $\lim_\lambda(u_\lambda + N_f) = \xi_f \in H_f$. Therefore

$$f(a) = \lim_\lambda f(u_\lambda a u_\lambda)$$

$$= \lim_\lambda (u_\lambda + N_f,\ au_\lambda + N_f)$$

$$= \lim_\lambda (u_\lambda + N_f,\ \pi_f(a)(u_\lambda + N_f))$$

$$=< \xi_f,\ \pi_f(a)\xi_f >,$$

for all $a \in A$.

Remark 1.31

(1) The above representation is unique in the sense of unitary equivalence.
(2) If we define $< a + N_f,\ b + N_f >= f(b^*a)$ for all $a,\ b \in A$,, then $f(a) =< \pi_f(a)\xi_f,\ \xi_f >$ for all $a \in A$.

F. Steinespring [26] extended the GNS construction to operator valued completely positive mapping and obtained the following representation theorem. For the concept of completely positive mapping and related conclusions, see the Appendix of this book.

In order to highlight the idea and avoid being trapped in the details, we consider only the case based on unital C^*-algebra as follows.

Proposition 1.15 *Suppose that A is a unital C^*-algebra and H is a Hilbert space. If $\phi : A \to B(H)$ is a completely positive mapping, then there exist Hilbert space K and a $*$ homomorphism$\pi : A \to B(K)$ preserving unit, and a bounded linear operator $V : H \to K$ satisfying $||\phi(1)|| = ||V||^2$, such that*

$$\phi(a) = V^*\pi(a)V,$$

for all $a \in A$.

Proof Define a sesquilinear form on $A \otimes_{alg} H$ which given on simple tensors by

$$A \otimes_{alg} H \times A \otimes_{alg} H \to \mathbb{C}, < a \otimes x,\ b \otimes y >=< x,\ \phi(a^*b)y > .$$

To check that the above form given in this way is nonnegative, the meaning of completely positive mapping is needed. For any $a_i \in A,\ x_i \in H,\ i = 1,\ 2,\ \cdots,\ n$, similar to the discussion in Sect. 1.3.1 the matrix $(a_i^*a_j)$ is a positive element in $M_n(A)$, it follows that $\phi_n((a_i^* a_j))$ is a positive element in $M_n(B(H)) = B(H^n)$ since ϕ is completely positive mapping. Hence we have

$$< \sum_{i=1}^{n} a_i \otimes x_i, \ \sum_{i=1}^{n} a_i \otimes x_i >$$

$$=< x, \ \phi_n((a_i^* a_j))x >\geq 0,$$

where $x = (x_1, \ \cdots, \ x_n) \in H^n$, which shows that the above sesquilinear form is nonnegative.

Let $N = \{u \in A \otimes_{alg} H :< u, \ u >= 0\}$. By Cauchy-Schwarz inequality

$$| < u, \ v > | \leq< u, \ u >< v, \ v > \ (u, \ v \in A \otimes_{alg} H),$$

then it is easy to check that $N = \{u \in A \otimes_{alg} H : < u, \ v >= 0 \text{ for all } v \in A \otimes_{alg} H\}$ is a closed subspace of $A \otimes_{alg} H$. So the quotient space $A \otimes_{alg} H/N$ is an inner product space given by

$$< u + N, \ v + N >=< u, \ v > \ (u, \ v \in A \otimes_{alg} H),$$

its completion is a Hilbert space, which denoted by K.

When $a \in A$. Define a linear mapping $\pi(a) : A \otimes_{alg} H \to A \otimes_{alg} H$ by the following way

$$\pi(a) \left(\sum_{i=1}^{n} a_i \otimes x_i \right) = \sum_{i=1}^{n} (aa_i) \otimes x_i, \ (a_i \in A, \ x_i \in H, i = 1, \ 2, \ \cdots, \ n).$$

Since $0 \leq (a_i^* a^* a a_j) \leq ||a^* a||(a_i^* a_j)$ by lemma 4.2 in [13], and ϕ_n is n-positive because ϕ is completely positive, it follows that $\phi_n((a_i^* a^* a a_j)) \leq ||a^* a||\phi_n((a_i^* a_j))$. Therefore

$$< \pi(a) \left(\sum_{i=1}^{n} a_i \otimes x_i \right), \ \pi(a) \left(\sum_{i=1}^{n} a_i \otimes x_i \right) >$$

$$= \sum_{i,j} < x_i, \ \phi(a_i^* a^* a a_j) >$$

$$=< x, \phi_n((a_i^* a^* a a_j))x >$$

$$\leq ||a||^2 < x, \ \phi_n((a_i^* a_j))x >$$

$$= ||a||^2 < \sum_{i=1}^{n} (a_i \otimes x_i), \sum_{i=1}^{n} (a_i \otimes x_i) >,$$

where $x = (x_1, \cdots, x_n)^T$. From the above proof, one can obtain that $||\pi(a)|| \leq ||a||$ and $\pi(a)N \subset N$. Thus $\pi(a)$ can be extended uniquely by continuity to a bounded linear operator on K, which is still denoted as $\pi(a)$. Routine calculation shows that $\pi : A \to B(K)$ is identity preserving $*$ homomorphism.

Define a linear operator $V : H \to K$ given by

$$V(x) = 1 \otimes x + N \ (x \in H).$$

Since

$$||Vx||^2 =< 1 \otimes x + N, \ 1 \otimes x + N >$$

$$=< 1 \otimes x, \ 1 \otimes x >=< x, \ \phi(1)x >$$

$$\leq ||\phi(1)||||x||^2.$$

This shows that V is bounded and $||V||^2 = \sup\{< x, \ \phi(1)x >: ||x|| \leq 1\} = \phi(1)$.

In the end, since

$$< V^*\pi(a)Vx, \ y >$$

$$=< \pi(a)(1 \otimes x + N), \ 1 \otimes y + N >$$

$$=< a \otimes x + N, \ 1 \otimes y + N >$$

$$=< x, \ \phi(a^*)y >=< \phi(a)x, \ y >,$$

for all $a \in A$ and $x, \ y \in H$. It implies that $\phi(a) = V^*\pi(a)V$ for all $a \in A$.

Remark 1.32

(1) The general case can see Theorem 3.6 in [27]. In particular, if the above Hilbert space H reduces to the complex field $\mathbb{C}$, then ϕ becomes a positive linear functional on C^*-algebra A. In this case, Proposition 1.15 becomes the GNS construction theorem.

(2) In Proposition 1.15, if $\phi(1) = 1$ also, then $V^*V = 1$, namely, V is an isometric operator. At this time, H can be taken as $VH(\subset K)$, and V^* is the projection from K onto H. Therefore,

$$\phi(a) = V^*\pi(a)|_H \ (a \in A).$$

(3) Moreover, if $T \in B(K)$, then $V^*T|H \in B(H)$ becomes a compressed mapping restricted to H. Let $K = H \oplus H^{\perp}$ be the orthogonal decomposition, then T can be regarded as an operator matrix of 2×2, so the compression of T on H is the $(1, 1)$ entry of this matrix. Thus, when $\phi(1) = 1$, Stinespring representation

theorem shows that every completely positive mapping from A to $B(H)$ is a compression corresponding to the $*$ homomorphism as described above;

(4) In Proposition 1.15, if letting K_1 be the linear closure of $\pi(A)VH$, then it is easy to check that K_1 reduces $\pi(A)$. Hence, $\pi_1 = \pi|_{K_1} : A \to B(K_1)$ is a $*$ homomorphism, and $\phi(a) = V^*\pi_1(a)V$. So (π_1, V, K_1) is also the Stinespring representation of ϕ. Moreover, this representation is unique in the sense of unitary equivalence, that is, if (π_2, V_2, K_2) is also a corresponding representation, there is a unitary operator $U : K_1 \to K_2$ such that $UV_1 = V_2, U\pi_1U^* = \pi_2$. In this case, (π_1, V, K_1) is called the minimal Stinespring representation of ϕ.

1.4.2 KSGNS Representation Theorem

Motivated by the above materials, Kasprove [11] extended the Stinespring representation theorem to Hilbert C^*-modules as follows.

Theorem 1.7 *Let E be a Hilbert C^*-module over a unital C^*-algebra A, and let $\phi : A \to L(E)$ be a completely positive mapping. Then the following conclusions hold:*

(1) There exists a Hilbert A-module E_ϕ, a $*$ homomorphism $\pi_\phi : A \to L(E_\phi)$ and a module operator $V_\phi \in L(E, E_\phi)$, such that

$$\phi(a) = V_\phi^*\pi_\phi(a)V_\phi \ (a \in A),$$

and $\pi_\phi(A)V_\phi E$ is dense in E_ϕ;

(2) The above representation is unique in the sense of unitary equivalence. That is, if (π, V, F) also meets the above requirements, then there exists a unitary operator $U \in L(E_\phi, F)$, such that

$$\pi(a) = U\pi_\phi(a)U^*, V = UV_\phi,$$

for all $a \in A$.

Triples $(\pi_\phi, V_\phi, E_\phi)$ is called the Kasparov-Stinespring-Gelfand-Naimark-Segal representation of the completely positive mapping ϕ, which is called the $KSGNS$ construction for short. Moreover, the representation satisfying the condition of Theorem 1.7 is called minimal representation.

Proof Define a A-valued sesquilinear form on $A \otimes_{alg} E$ given by

$$< a \otimes x_1, b \otimes x_2 >=< x_1, \phi(a^*b)x_2 >,$$

for all $x_1,\ x_2 \in E$ and $a,\ b \in A$. Since

$$< \sum_{i=1}^{n} a_i \otimes x_i,\ \sum_{i=1}^{n} a_i \otimes x_i >$$

$$= \sum_{i,j}^{n} < x_i,\ \phi(a_i^* a_j) x_j > = < x,\ \phi_n(a) x > \geq 0,$$

where $x = (x_1,\ \cdots,\ x_n)^T$ and $a = (a_i^* a_j) \in M_n^+(A)$, which shows that the above form is a non-negative A-valued inner product.

Let $N_\phi = \{z \in A \otimes_{alg} E :< z,\ z >= 0\}$, then quotient space $A \otimes_{alg} E/N_\phi$ is a pre-Hilbert A-module, so that the completion of $A \otimes_{alg} E/N_\phi$ is a Hilbert A-module, which is denoted by E_ϕ. Similar to the proof of the corresponding part of Proposition 1.15, we can define a $*$ homomorphism $\pi_\phi : A \to L(E_\phi)$ which preserves the identity, such that for any $a \in A$ and any $a' \otimes x \in A \otimes_{alg} E$ we have

$$\pi_\phi(a)(a' \otimes x + N_\phi) = aa' \otimes x + N_\phi.$$

Define a A-linear operator $V_\phi : E \to E_\phi$ given by

$$V_\phi(x) = 1 \otimes x + N_\phi \ (x \in E).$$

We want to prove that $V_\phi \in L(E,\ E_\phi)$. For this, let $W_\phi :\ A \otimes_{alg} E/N_\phi \to E$ by $a \otimes y + N_\phi \to \phi(a)y$ for all $a \in A$ and $y \in E$. Since for all $a_i \in A,\ y_i \in E,\ i = 1,\ \cdots,\ n$, we have

$$||\sum_{i=1}^{n} \phi(a_i) y_i||^2 = ||\sum_{i,j}^{n} < y_i, \phi(a_i^*)\phi(a_j) y_j > ||$$

$$\leq ||\phi|| \cdot ||\sum_{i,j}^{n} < y_i,\ \phi(a_i^* a_j) y_j > ||$$

$$= ||\phi|| \cdot ||\sum_{i=1}^{n} (a_i \otimes y_i) + N_\phi||^2,$$

by lemma 5.4 in [13]. Hence W_ϕ can be extended a bounded A-linear mapping from E_ϕ to E. Since

$$< V_\phi(x),\ a \otimes y + N_\phi >$$

$$=< 1 \otimes x + N_\phi,\ a \otimes y + N_\phi >$$

$$=< 1 \otimes x,\ a \otimes y >=< x,\ \phi(a)y >$$

$$=< x,\ W_\phi(a \otimes y + N_\phi) >,$$

which implies that $V_\phi \in L(E, E_\phi)$ and $W_\phi = V_\phi^*$. Therefore for any $a \in$ and $x \in E$, we have

$$V_\phi^* \pi_\phi(a) V_\phi = V_\phi^* \pi_\phi(a)(1 \otimes x + N_\phi).$$

$$= V_\phi^*(a \otimes x + N_\phi) = \phi(a)x$$

Hence $\phi(a) = V_\phi^* \pi_\phi V_\phi$. Note that $1 \in A$, so that $\pi_\phi(A) V_\phi E$ is dense in E_ϕ.

Finally, we have to prove the uniqueness of the above representation in the sense of isometric isomorphism. If the triplet (π, V, F) also meets the requirements of Theorem 1.7, then for any given element $a_i \in A,\ x_i \in E,\ i = 1,\ 2, \cdots,\ n$, we have

$$||\sum_{i=1}^{n} \pi(a_i) V x_i||^2 = ||\sum_{i,j}^{n} < x_i,\ V^* \pi(a_i^* a_j) V x_j > ||$$

$$= ||\sum_{i,j}^{n} < x_i,\ \phi(a_i^* a_j) y_j > ||$$

$$= ||\sum_{i,j}^{n} < x_i,\ V_\phi^* \pi_\phi(a_i^* a_j) V_\phi x_j > ||$$

$$= ||\sum_{i=1}^{n} \pi_\phi(a_i) V_\phi x_i||^2.$$

Hence $U : \sum_{i=1}^{n} \pi_\phi(a_i)V_\phi x_i \to \sum_{i=1}^{n} \pi(a_i)Vx_i$ can be extended to an isometric module mapping from E_ϕ onto F, which is still denoted by U. Thus U is a unitary operator. Furthermore, since

$$\begin{aligned} U\pi_\phi(a)U^* \left(\sum_{i=1}^{n} \pi(a_i)Vx_i\right) \\ = U\pi_\phi(a)\left(\sum_{i=1}^{n} \pi_\phi(a_i)V_\phi x_i\right) \\ = U\left(\sum_{i=1}^{n} \pi_\phi(aa_i)V_\phi x_i\right) \\ = \sum_{i=1}^{n} \pi(aa_i)Vx_i \\ = \pi(a)\left(\sum_{i=1}^{n} \pi(a_i)Vx_i\right), \end{aligned}$$

from which it implies that $U\pi_\phi(a)U^* = \pi(a)$. On the other hand, by the construction of U, we have

$$\begin{aligned} UV_\phi x &= U\pi_\phi(1)V_\phi x \\ &= \pi(1)Vx = Vx, \end{aligned}$$

for all $x \in E$. So that $UV_\phi = V$.

Remark 1.33

(1) If A has no unit, then we need to add the condition that ϕ is strictly continuous, that is, $\{\phi(u_\lambda)\}$ is a Cauchy net in strict topology in $L(E)$ for some approximation unit $\{u_\lambda\}$ of A. In fact, the strict continuity of ϕ is equivalent to nondegeneracy of π_ϕ (see Proposition 5.5 and Theorem 5.6 in [13] for details). In particular, if ϕ is a $*$ homomorphism, then ϕ is strictly continuous if and only if the closed submodule $\overline{\phi(A)E}$ is complemented in E (see Proposition 5.8 in [13]).

(2) In Theorem 1.7, if $A = E = \mathbb{C}$,, then the $KSGNS$ construction becomes the classical GNS construction; if E reduces to Hilbert space, then $KSGNS$ construction becomes Stinespring representation theorem.

References

1. Alegra Dajić, Koliha, J.J.: Positive solutions to the equations and for Hilbert space operators. J. Math. Anal. Appl. **333**(2), 567–576 (2007)
2. Chang-Zhou Dong, Qing-Wen Wang, Yu-Ping Zhang: The common positive solution to adjointable operator equations. J. Math. Anal. Appl. **396**, 670–679 (2012)
3. Cvetković-Ilić, D., Qing-Wen Wang, Qingxiang Xu, Douglas'+Sebesty é n's lemmas=a tool for solving an operator equation problem. J. Math. Anal. Appl. **482**, 123599 (2020)
4. Douglas, R.G.: On majorization, factorization, and range inclusion of operators on Hilbert space. Proc. AMS. **17**, 413–415 (1966)
5. Farid, F.D., Moslehian, M.S., Qing-Wen Wang, Zong-Cheng Wu: On the Hermitian solutions to a system of adjointable operator equations. Linear Algebra Appl. **437**, 1854–1891 (2012)
6. Frank, M.: Self-duality and C^*-reflexivity of Hilbert C^*-moduli. Z. Anal. Anwendungen **9**, 165–176 (1990)
7. Frank, M.: Generalized inverses and polar decomposition of unbounded regular operators on Hilbert C^*-modules. J. Oper. Theory **64**(2), 377–386 (2010)
8. Harte, R.E., Mbekhta, M.: On generalized inverses in C^*-algebras, I. Stud. Math. **103**, 71–77 (1992)
9. Huaxin Lin: Hilbert C^*-modules and bounded module mappings. Scientia Sinica (Mathematica) **12**, 1243–1252 (1990)
10. Huaxin Lin: Bounded module maps and pure completely positive maps. J. Oper. Theory **26**, 121–138 (1991)
11. Kasparov, G.G.: Hilbert C^*-modules: theorems of Stinespring and Voiculescu. J. Oper. Theory **4**, 133–150 (1980)
12. Kusuda, M.: Characterizations of certain classes of hereditary C^*-subalgebras. Proc. AMS. **116**(4), 999–1005 (1992)
13. Lance, E.C.: Hilbert C*-Modules: A Toolkit for Operator Algebraists, London Mathematical Society. Lecture Note Series, vol. 124. Cambridge University Press, Cambridge (1994)
14. Lance, E.C.: Unitary operator on Hilbert C^*-modules. Bull. Lond. Math. Soc. **26**, 363–366 (1994)
15. Li Bengren: Operators Algebras. Basic Modern Mathematics Series. China Science Press, Beijing (1986)
16. Magajna, B.: Hilbert C^*-modules in which all closed submodules are complemented. Proc. AMS. **125**(3), 849–852 (1997)
17. Murphy, G.J.: C^*-Algebras and Operator Theory. Academic, London (1990)
18. Paschke, W.L.: Inner product modules overB^*-algebras. Trans. Am. Math. Soc. **182**, 443–468 (1973)
19. Popovici, D.: Orthogonal decompositions of isometries in Hilbert C^*-modules. J. Oper. Theory **39**, 99–112 (1998)
20. Qing-Wen Wang, Chang-Zhou Dong: The positive solution to a system of adjointable operator equations over Hilbert C^*-modules. Linear Algebra Appl. **433**, 1481–1489 (2010)
21. Qingxiang Xu: Common Hermitian and positive solutions to the adjointable operator equations $AX = C, XB = D$. Linear Algebra Appl. **429**, 1–11 (2008)
22. Qingxiang Xu, Lijuan Sheng: Positive semi-definite matrices of adjointable operators on Hilbert C^*-modules. Linear Algebra Appl. **428**(4), 992–1000 (2008)
23. Qingxiang Xu, Lijuan Sheng, Yangyang Gu: The solutions to some operator equations. Linear Algebra Appl. **429**, 1997–C2024 (2008)
24. Schweizer, J.: Hilbert C^*-modules with a predual. J. Oper. Theory **48**, 621–632 (2002)
25. Skeide, M.: Generalised matrix C^*-algebras and representations of Hilbert modules, Mathematical Proceedings, Royal Irish Academy **100**, 11–38 (2000)
26. Stinespring, F.: Positive functions on C^*-algebras. Proc. AMS. **6**, 211–216 (1955)
27. Takesaki, M.: Theory of Operator Algebras I. Springer, New York/Berlin/Heidelberg (1979)

28. Wenting Liang, Chunyuan Deng: The solutions to some operator equations with corresponding operators not necessarily having closed ranges. Linear Multilinear Algebra **67**(8), 1606–1624 (2019)
29. Xiaochun Fang, Jing Yu: Solutions to operator equations on Hilbert $C*$-modules II. Integral Equ. Oper. Theory **68**, 23–60 (2010)
30. Zhang Lunchuan: The characterization of bounded generalized inverse module maps and applying in C^*-algebras. Abstracts of Short Communications and Poster Sessions. 9 Operator Algebras and Functional Analysis, P_{164}, ICM2002. Higher Education Press, Beijing (2002)
31. Zhang Lunchuan: Relation between Hereditary C^*-subalgebras and complemented submodules. Acta Math. Sinica (Chinese) **47**(4), 747–750 (2004)
32. Zhang Lunchuan: The factor decomposition theorem of bounded generalized inverse module maps. Acta Math. Sinica English Ser. **23**(8), 1413–1418 (2007)

Chapter 2
Kasprove's Stabilization and Fredholm Generalized Index Theory

In this chapter, we show that every countably generated Hilbert A-module can be regarded as a complemented submodule of $H \otimes A$ (in the sense of unitary equivalence), where H is a separable Hilbert space. This is one of the fundamental results of Hilbert C^*-module theory, which is called Kasprove's stabilization theorem. As one of the applications of Hilbert A-module theory, we characterize Morita equivalence between C^*-algebras and prove that two σ-*unital* C^* algebras are stable isomorphism if and only if they are Morita equivalent. Then we go through Fredholm generalized index theory based on $H \otimes A$; the connection to index map in K-theory is discussed. In the final section of this chapter, we introduce the elementary theory of module frames. The relationship between unitary equivalence of closed submodules and stable isomorphism of the corresponding hereditary C^*-subalgebras will be characterized by using the module framework as tool.

2.1 Kasprove's Stabilization Theorem

2.1.1 σ-*unital* C^*-Algebras

In order to characterize the Kasprove's stabilization theorem, we need the basic knowledge of σ-*unital* C^* algebra. Recall that a C^*-algebra is said to be σ-*unital* if it has a countable approximate unit. Therefore, each separable C^*-algebra is σ-*unital* C^*-algebra.

Definition 2.1 Let A be a C^*-algebra and a be in A^+, where A^+ is the set of all positive elements in A. Then a is called strictly positive element if $\tau(a) > 0$ for every state τ on A.

L. Zhang, *Hilbert C*- Modules and Quantum Markov Semigroups*,
https://doi.org/10.1007/978-981-99-8668-2_2

The following is an equivalent characterization of strictly positive elements:

Proposition 2.1 *Suppose that A is a C^*-algebra and a is in A. Then the following conditions are equivalent:*

(1) a is a strictly positive element;
(2) The hereditary C^*-subalgebra $\overline{aAa}$ generated by a is A;
(3) If A is unital, then (1) *and* (2) *are also equivalent to that a is invertible.*

Proof (1) $\Rightarrow$ (2) : Assume that $\overline{aAa} \neq A$. Then there exists a pure state ρ on A such that $\rho(\overline{aAa}) = 0$, which follows that $\rho(a) = 0$, since $a \in \overline{aAa}$ from the basic properties of positive elements of C^*-algebras. This is in contradiction with condition (1).

(2) $\Rightarrow$ (1) : Assume that there exists a state ρ on A such that $\rho(a) = 0$. Then for each $x \in A$,, we have

$$|\rho(axa)|^2 = |\rho(a^{\frac{1}{2}}(a^{\frac{1}{2}}xa))|^2$$

$$\leq |\rho(a)||\rho((a^{\frac{1}{2}}xa)^*(a^{\frac{1}{2}}xa))| = 0,$$

by Cauchy-Schwarz inequality, from which it implies that $\rho(\overline{aAa}) = 0$, equivalently $\overline{aAa} \neq A$. This leads to contradiction with condition (2).

In addition, when A is unital, we shall prove that (1) $\Leftrightarrow$ (3). First, to show that (1) $\Rightarrow$ (3). Denote by $C^*(a)$ the commutative C^*-subalgebra of A which is generated by a and unit 1. It results from Gelfand representation theorem that $C^*(a) \cong C(\Omega)$, where Ω is the character space of $C^*(a)$ which is the set of all nonzero $*$ homomorphisms from $C^*(a)$ to the complex field $\mathbb{C}$. For any fixed $t \in \Omega$, since $f(h) = h(t)$ $(h \in C(\Omega))$, is a state on $C(\Omega)$, and since state on $C^*(a)$ can be extended to the state on A. It follows from the Gelfand transform and combing with the condition (1) that

$$f(\hat{a}) = \hat{a}(t) = t(a) > 0,$$

where $\hat{a} \in C(\Omega) : \hat{a}(t) = t(a)$ $(t \in \Omega)$, is the Gelfand transform of a. This shows that $0 \notin \sigma(a)$ since the spectrum $\sigma(a)$ of a is equal to $\hat{a}(\Omega)$. Furthermore, since $\{0\}$ is closed, and since $\sigma(a)$ is compact in the complex field $\mathbb{C}$, there exists a $\lambda > 0$ such that $\sigma(a) \subset [\lambda, +\infty)$, which shows that a is reversible.

Next, to show that (3) $\Rightarrow$ (1). Since the positive element a is reversible, there exists a positive number λ such that $a \geq \lambda 1$, where 1 is the unit of A. Therefore $f(a - \lambda 1) \geq 0$ for all nonzero positive linear functional f on A. Thus

$$f(a) \geq \lambda f(1) = \lambda||f|| > 0.$$

Remark 2.1 According to the one-to-one correspondence between the set of closed right ideals and the set of hereditary C^*-subalgebras, from Proposition 2.1, we can get the following conclusion: Given a unital C^*-algebra A and a positive element a, then a is a strictly positive element if and only if the closed right ideal $\overline{aA}$ generated by a is A.

In the following, an equivalent characterization of σ-unital C^*-algebra is given.

Proposition 2.2 *Suppose that A is a C^*-algebra. Then A is σ-unital, if and only if A has strictly positive element.*

Proof If A is σ-*unital*, then there exists countable approximation unit $\{a_n\}_{n=1}^{\infty} \subset A$. Put $a = \sum_{n=1}^{\infty} \frac{1}{2^n} a_n$, then a is a strictly positive element in A. Suppose contrary, there exists a state f on A such that $f(a) = 0$, that is,

$$\sum_{n=1}^{\infty} \frac{1}{2^n} f(a_n) = 0.$$

Hence $f(a_n) = 0$ $(n \in \mathbb{N})$ since each $f(a_n) \geq 0$ $(n \in \mathbb{N})$. Therefore, for any $x \in A$, we can obtain

$$|f(x)|^2 = \lim_{n \to +\infty} |f(a_n x)|^2$$

$$= \lim_{n \to +\infty} |f(a_n^{\frac{1}{2}} a_n^{\frac{1}{2}} x)|^2$$

$$\leq \lim_{n \to +\infty} |f(a_n)||f((a_n^{\frac{1}{2}} x)^*(a_n^{\frac{1}{2}} x))| = 0.$$

This is in contradiction with $f \neq 0$.

Conversely, suppose that a is a strictly positive element in A. Let $g_n(t) = t/(t + \frac{1}{n})$, $t \in \sigma(a)$. It is easy to check that $\{g_n(t)\}_{n=1}^{\infty}$ is an operator-monotone function sequence. Using the function calculus of positive elements of C^*-algebra, let

$$a_n = a\left(a + \frac{1}{n}\right)^{-1}.$$

By direct calculus, we see that $\{a_n\}_{n=1}^{\infty}$ is an increasing approximation unit in A.

2.1.2 Kasprove's Stabilization Theorem

Theorem 2.1 (Kasprove's stabilization theorem) *Let E be a countably generated Hilbert C^*-module over a C^*-algebra A, and let H be a separable Hilbert space. Then*

$$E \oplus (H \otimes A) \cong H \otimes A.$$

We first give a lemma.

Lemma 2.1 *Suppose that E is a Hilbert A-module and T is a positive element in $K(E)$. Then T is strictly positive in $K(E)$ if and only if $\overline{R(T)} = E$.*

Proof If T is a strictly positive element in $K(E)$, then $\overline{TK(E)} = K(E)$ by Remark 2.1. Hence $\overline{TE} = E$ since $\overline{K(E)E} = E$.

Conversely, suppose that $\overline{TE} = E$. Then for any $x \in E$, there exists sequence $\{x_n\}$ in E such that $Tx_n \to x,\ n \to +\infty$ (convergence in norm in E). It follows that

$$\theta_{x,\,y} = \lim_{n\to+\infty} \theta_{Tx_n,\,y}$$

$$= \lim_{n\to+\infty} T\theta_{x_n,\,y}\ (y \in E) \in \overline{TK(E)}.$$

Therefore $K(E) \subset \overline{TK(E)} \subset K(E)$, since $K(E)$ is generated by elements of the form $\theta_{x,\,y}$. The above proof shows that $\overline{TK(E)} = K(E)$. Thus T is a strictly positive element from Remark 2.1 again.

The proof of Theorem 2.1. First, consider the case that A has unit 1. We write $\{e_n\}_{n=1}^{\infty}$ for the generating set of E. Set $e_{pq} = e_q,\ p = 1, 2, \cdots; q = 1, 2, \cdots$. Construct a sequence $\{y_n\} \subset E$ by the following way:

If $p = q = 1$, then let $y_1 = e_{11} = e_1$. If $p+q > 2$, then let $y_n = e_{pq} = e_q$, where $n = q+\sum_{k=1}^{p+q-2} k$. Hence $\{y_n\} \subset E$ and in which each element of the generating set occurs infinitely by the construction of $\{y_n\}$. Take $\{\varepsilon_n\}$ to be a standard orthogonal basis of H, then $\{\varepsilon_n \otimes 1\}$ is a standard orthogonal basis of $H \otimes A$, write $f_n = \varepsilon_n \otimes 1$ for convenience. Define a module operator $T \in K(H \otimes A,\ E \oplus (H \otimes A))$ by

$$T = \sum_{i=1}^{\infty} \frac{1}{2^i}\theta_{y_i, f_i} + \sum_{i=1}^{\infty} \frac{1}{4^i}\theta_{f_i,\, f_i}.$$

We prove next that $\overline{R(T)} = E \oplus (H \otimes A)$. Indeed, for every $f_m \in \{f_n\}$, we have

$$T(2^m f_m) = y_m + \frac{1}{2^m} f_m \in \overline{R(T)}.$$

From the construction of $\{y_n\}$, there exists a subsequence n_m of natural number set $\mathbb{N}$ associated with a fixed natural number n, such that $y_{n_m} = y_n$. It follows that

$$y_n \oplus \frac{1}{2^{n_m}} f_{n_m} = y_{n_m} \oplus \frac{1}{2^{n_m}} f_{n_m} \in \overline{R(T)}.$$

So that $(y_n,\ 0) \in \overline{R(T)}$ when $m \to +\infty$. Since $T(4^n f_n) = 2^n y_n \oplus f_n$,, we have then

$$(0,\ f_n) = T(4^n f_n) - (2^n y_n,\ 0) \in \overline{R(T)},$$

which shows that $\overline{R(T)} = E \oplus (H \otimes A)$. Note that $T^* = \sum_{i=1}^{\infty} \frac{1}{2^i}\theta_{f_i,\ y_i} + \sum_{i=1}^{\infty} \frac{1}{4^i}\theta_{f_i,\ f_i}$. Therefore

$$T^*T = \left(\sum_{i=1}^{\infty} \frac{1}{2^i}\theta_{y_i,\ f_i}\right)^* \left(\sum_{i=1}^{\infty} \frac{1}{2^i}\theta_{y_i,\ f_i}\right) + \sum_{i=1}^{\infty} \frac{1}{4^{2i}}\theta_{f_i,\ f_i}$$

$$\geq \sum_{i=1}^{\infty} \frac{1}{4^{2i}}\theta_{f_i,\ f_i} > 0.$$

It is easy to check that $\sum_{i=1}^{\infty} \frac{1}{4^{2i}}\theta_{f_i,\ f_i}$ has dense range in $H \otimes A$. Hence $\sum_{i=1}^{\infty} \frac{1}{4^{2i}}\theta_{f_i,\ f_i}$ is a strictly positive element by Lemma 2.1. Thus T^*T is strictly positive and $\overline{R(T^*T)} = H \otimes A$ by Lemma 2.1 again. It follows that $|T| = (T^*T)^{1/2}$ has also dense range in $H \otimes A$. Furthermore, define a A-linear mapping $U : H \otimes A \to E \oplus (H \otimes A)$ by $U(|T|x) = Tx\ (x \in H \otimes A)$. Since

$$< U(|T|x),\ U(|T|y) >=< Tx,\ Ty >$$

$$=< x,\ T^*Ty >=< |T|x,\ |T|y >$$

for all $x,\ y \in H \otimes A$, U can be extended to an isometric surjection from $H \otimes A$ onto $E \oplus (H \otimes A)$, namely, U is unitary.

Finally, suppose that A is not unital. Let $\widetilde{A} = A \oplus \mathbb{C}$, setting $x \cdot 1 = x\ (x \in E)$, in which 1 stands for the unit $(0, 1)$ of $\widetilde{A}$. Therefore E becomes a Hilbert C^*-module over the unital C^*-algebra $\widetilde{A}$. It is easy to check that module mapping $U : (\xi \otimes 1)a \to \xi \otimes a\ (\xi \in H,\ a \in A)$ can be extended to unitary operator from $\overline{(H \otimes \widetilde{A})A}$ onto $H \otimes A$. Similarly, $V : xa + (\xi \otimes 1)a \to xa + \xi \otimes a\ (\xi \in H,\ a \in A)$ can be extended to unitary module operator from $\overline{(E \oplus (H \otimes \widetilde{A}))A}$ onto $E \oplus (H \otimes A)$. So it is equivalently to prove the following claim:

$$E \oplus (H \otimes \widetilde{A}) \cong H \otimes \widetilde{A}.$$

This is exactly what we have proved in the first part, that is, the case without unit can be transformed into the case with unit. We are done.

Remark 2.2 In the proof of Theorem 2.1, we have used the fact that the cross term is zero in the expression of T^*T. Since E and $H \otimes A$ are complemented in $E \oplus (H \otimes A)$, hence identifying E with $(E,\ 0)$ and $H \otimes A$ with $(0,\ H \otimes A)$, respectively. Therefore $< y_n,\ f_m >=< f_m,\ y_n >= 0$.

The classification of Hilbert C^*-modules under unitary equivalence is almost impossible for general Hilbert A-modules. But for countably generated Hilbert A-modules, Kasprove's stabilization theorem provides a method. For this purpose, in what follows the concepts fully complemented submodules and fully complemented projections are introduced.

Suppose that E is a Hilbert A-module and closed submodule F in it. If F is complemented in E and $F^{\perp} \cong E$, then F is called a fully complemented submodule. Furthermore, a project element p in $L(E)$ is said to be fully complemented if its range $R(p)$ is a fully complemented submodule in E. The following consequence is obtained immediately from the Kasprove's stabilization theorem.

Corollary 2.1 *Suppose that E is countably generated Hilbert A-module. Then there exists a fully complemented submodule F in $H \otimes A$ such that E is unitary equivalent to F, namely, $E \cong F$, where H is a separable Hilbert space.*

Indeed, applying Kasprove's stabilization theorem, there exists unitary module operator $U : E \oplus (H \otimes A) \to H \otimes A$. Hence UE and $U(H \otimes A)$ are closed submodules in $H \otimes A$, and we have then

$$E \cong UE; \ (UE)^{\perp} = U(H \otimes A) \cong H \otimes A.$$

Thus, UE meets the requirement.

To sum up, the classification problem of countably generated Hilbert C^*-A-modules is equivalent to the classification problem of fully complemented projective elements in $L(H \otimes A)$.

At the end of this subsection, by using the Kasprove's stabilization theorem as a tool, we give the relation between the countably generated Hilbert A-module and $\sigma\text{-}unital$ C^*-algebras. For this, we shall start from Hilbert space to look for the trail.

Proposition 2.3 *If H is a Hilbert space. Then H is separable if and only if $K(H)$ is $\sigma\text{-}unital$ C^*-algebra.*

Proof Suppose that H is a separable Hilbert space. Let $\{e_n\}_{n=1}^{\infty}$ be an standard orthogonal basis of H, and we write H_n for the closed subspace of H which is generated by $\{e_1, \ e_2, \ \cdots, \ e_n\}$. Define projection operator $P_n : H \to H_n$, then it is easy to check that monotone increasing sequence $\{P_n\}_{n=1}^{\infty} \subset K(H)$ is an approximate unit of $K(H)$. Thus $K(H)$ is a $\sigma\text{-}unital$ C^*-algebra.

Conversely, suppose that $K(H)$ is a $\sigma\text{-}unital$ C^*-algebra. Let $\{u_n\}_{n=1}^{\infty}$ be an approximate unit of $K(H)$. Since $F(H)$ consisting of all finite rank operators is dense in $K(H)$. For any fixed $n \in \mathbb{N}$, there exists a sequence $x_{n_i}, \ y_{n_i} \in H, \ i = 1, \ 2, \ \cdots, \ m(n)$, where the natural number $m(n)$ is depending on n, such that

$$||\sum_{i=1}^{m(n)} \theta_{x_{n_i}, \ y_{n_i}} - u_n|| < \frac{1}{n}.$$

Since approximate unit of $K(H)$ converges to the identity I in $B(H)$ with respect to the strong $*$-topology, from which it shows that for any $x \in H$, we have $u_n x \to x,\ n \to +\infty$. Therefore

$$||\sum_{i=1}^{m(n)} \theta_{x_{n_i},\ y_{n_i}} x - x|| \to 0,\ n \to \infty.$$

Thus H is generated by the countable set $\{x_{n_i},\ 1 \leq n_i \leq m(n),\ n \in \mathbb{N}\}$ since $\theta_{x_{n_i},\ y_{n_i}} x = x_{n_i} < y_{n_i},\ x > .$

Generalizing Proposition 2.3 to the framework of Hilbert A-modules, the following conclusions can be obtained.

Proposition 2.4 *Hilbert A-module E is countably generated if and only if C^*-algebra $K(E)$ is σ-unital.*

Proof Without loss of generality, suppose that A is unital from the proof of Theorem 2.1. If E countably generated, then by Theorem 2.1, there exists a Hilbert A-module F such that $E \oplus F \cong H \otimes A$. We regard E as a complemented submodule in $H \otimes A$ (in the sense of unitary equivalence), then there exists a projection mapping $P_E : H \otimes A \to E$ in $L(H \otimes A, E)$. Let $\{e_n\}_{n=1}^{\infty}$ be a standard orthogonal basis of $H \otimes A$, then $\{\sum_{i=1}^{n} \frac{1}{2^i} \theta_{e_i, e_i}\}_{n=1}^{\infty}$ is a strictly positive element of $K(H \otimes A)$ by the proof of Theorem 2.1. Therefore, there exists a countable approximation unit $\{u_n\}_{n=1}^{\infty}$ in $K(H \otimes A)$ by Proposition 2.1. It follows by direct calculus that $\{P_E u_n P_E^*\}_{n=1}^{\infty}$ is an approximate unit of $K(E)$. Thus, $K(E)$ is σ-*unital*.

Conversely, if $K(E)$ is σ-*unital*, let $\{u_n\}_{n=1}^{\infty}$ is an approximate unit of $K(E)$. The following is parallel to the proof of Proposition 2.3, so that E is countable generated.

2.2 Morita Equivalence and C^*-Correspondences

2.2.1 Morita Equivalence Theorem

This section is one of the applications of Hilbert C^*-module theory. Morita equivalence is the main tool for characterizing the stable isomorphism between C^*-algebras. It is known that the C^*-algebras which are stable isomorphism have the same $K(K_0$ and $K_1)$-groups. Therefore, it is of great significance to describe Morita equivalence.

Recall that a C^*-algebra A is called stable if $A \cong A \otimes K(H)$ and C^*-algebras A and B are called stable isomorphism if $A \otimes K(H) \cong B \otimes K(H)$, where H is a separable Hilbert space.

The concept of equivalent bimodule is introduced as follows.

Definition 2.2 Let A and B be C^*-algebras, and let linear space E be right A-module and left B-module, simultaneously. Then E is called a Hilbert A-B-bimodule if it is a Hilbert A-module and is simultaneously a Hilbert B-module such that

$$\xi \langle \eta,\ \tau \rangle_A =_B \langle \xi, \eta \rangle \tau \ (\xi,\ \eta,\ \tau \in E).$$

Furthermore, if E is full as a right A-module and left B-module, respectively, that is, $\overline{\langle E, E \rangle_A} = A$ and $\overline{_B\langle E, E \rangle} = B$,, then E is called equivalence A-B-bimodule, or equivalence bimodule in short.

C^*-algebras A and B are called Morita equivalence in case there exists an equivalence A-B-bimodule, which is denoted by $A \sim_M B$.

We first give a necessary and sufficient condition for Morita equivalence.

Theorem 2.2 *Let A and B be C^*-algebras. Then the following conditions are equivalent:*

(1) $A \sim_M B$;
(2) There exists a full Hilbert right A-module F such that $K_A(F) \cong B$.

Lemma 2.2 *If E is a Hilbert A-module, then $\overline{< E,\ E >}$ has an approximation unit $\{u_\alpha\}_{\alpha \in \Gamma}$ with the following properties: Let Γ be a set made up of all finite subsets of E, which is a directed set with respect to the inclusion relation of sets. Each of them has the form $u_\alpha = \sum_{i=1}^{n(\alpha)} < \eta_i^{(\alpha)},\ \eta_i^{(\alpha)} > \ (\alpha \in \Gamma)$, where $\eta_i^{(\alpha)} \in E$ is determined by α, and $n(\alpha)$ is the number of elements contained in α.*

Proof The following construction is derived from Dixmier [8] 1.7.2. Recall that Γ is the set of all finite subsets of E, and it is a directed set with the inclusion relation which is denoted by "$\subseteq$." For any $\alpha = \{\xi_1,\ \xi_2, \cdots,\ \xi_{n(\alpha)}\} \in \Gamma$, let

$$\eta_i^{(\alpha)} = \xi_i \left(n(\alpha)^{-1} + \sum_{i=1}^{n(\alpha)} < \xi_i,\ \xi_i > \right)^{-1/2},$$

where $(n(\alpha)^{-1} + \sum_{i=1}^{n(\alpha)} < \xi_i,\ \xi_i >)^{-1/2} \in \widetilde{A} = A \oplus \mathbb{C}$. Then using it to construct u_α given by

$$u_\alpha = \sum_{i=1}^{n(\alpha)} < \eta_i^{(\alpha)},\ \eta_i^{(\alpha)} >,$$

from this, we get a sequence $\{u_\alpha\}_{\alpha \in \Gamma}$. We prove next that $\{u_\alpha\}$ is an approximate unit of $\overline{< E,\ E >}$. Since any element in $\overline{< E,\ E >}$ can be approximated by a sequence consisting of finite linear combinations of the form $< \xi,\ \eta >$, we may assume that $\xi = \eta$ by polarization. Let $\Gamma_\xi \subset \Gamma$ be the set made up of all subsets containing ξ. Then by the function calculus of the positive element of C^*-algebra, we have

$$||u_\alpha|| \leq 1,$$

for all $\alpha \in \Gamma$, and

$$||u_\lambda < \xi,\ \xi > - < \xi,\ \xi > || \to 0,\ \lambda \in F_\xi.$$

Thus $\{u_\alpha\}$ is an approximate unit of $\overline{< E,\ E >}$.

Corollary 2.2 *If A is a σ-unital C^*-algebra and E is a full Hilbert C^*-A-module, then there exists a sequence $\{x_n\} \subset E$, such that $\{\sum_{i=1}^n < x_i,\ x_i >\}_{n=1}^\infty$ is an approximate unit of A.*

Proof Since A is σ-*unital*, there is a countable approximate unit $\{a_n\}$ in A. Because E is full, that is, $\overline{< E,\ E >} = A$, so by Lemma 2.2, there exists an approximate unit $\{u_\alpha\}_{\alpha\in\Gamma}$ in A which has the following form: Each $u_\alpha = \sum_{i=1}^{n(\alpha)} < \eta_i^{(\alpha)}, \eta_i^{(\alpha)} >$ $(\eta_i^{(\alpha)} \in E,\ i = 1,\ 2,\ \cdots,\ n(\alpha))$. Here we rewrite the relation "$\subseteq$" as "$\leq$." So, for a_1, there exists $\alpha_1 \in \Gamma$ such that $||a_1 u_\alpha a_1 - a_1|| < 1$ when $\alpha \geq \alpha_1$. For a_2 there exists $\alpha_2 \in \Gamma$ and $\alpha_2 \geq \alpha_1$, such that $||a_2 u_\alpha a_2 - a_2|| < \frac{1}{2}$ when $\alpha \geq \alpha_2$. Continue this process, for a_n there exists $\alpha_n \in \Gamma$ and $\alpha_n \geq \alpha_{n-1}$, such that

$$||a_n u_\alpha a_n - a_n|| < \frac{1}{n}\ (n \geq 2,\ n \in \mathbb{N}),$$

when $\alpha \geq \alpha_n$. Therefore $\{a_n u_{\alpha_n} a_n\}$ is a countable approximate unit in A. Since $a_n u_{\alpha_n} a_n = \sum_{i=1}^{n(\alpha_n)} < \eta_i^{(\alpha_n)} a_n,\ \eta_i^{(\alpha_n)} a_n >$, renumber $\eta_i^{(\alpha_n)} a_n$ corresponding to $(\alpha_n)_{n=1}^\infty$ as $(x_i)_{i=1}^\infty$, it is easy to check that $\{\sum_{i=1}^n < x_i,\ x_i >\}_{n=1}^\infty$ is an approximate unit in A.

Remark 2.3 According to Lemma 2.2, if E is an equivalent A-B-bimodule, then the following equalities hold:

$$\langle b\eta,\ \mu\rangle_A = \langle \eta,\ b^*\mu\rangle_A,$$

and

$$_B\langle \eta a, \mu\rangle =_B \langle \eta, \mu a^*\rangle,$$

for any $\eta,\ \mu \in E$ and $a \in A,\ b \in B$.

We first prove the following claim: If c in $\overline{< E,\ E >_A}$ and $\xi c = 0$ for all $\xi \in E$, then $c = 0$.

Indeed, $c^* < \xi,\ \xi >_A c =< \xi c,\ \xi c >_A = \underline{0 \text{ since } \xi c = 0}$. It follows by Lemma 2.2 that for any approximate unit $\{u_\lambda\}_{\lambda\in\Gamma}$ in $\overline{< E,\ E >_A}$, we have $c^* u_\lambda c = 0\ (\lambda \in \Gamma)$. It implies that $c^* c = \lim_\lambda c^* u_\lambda c = 0$, so $c = 0$.

Now we prove the conclusion. Since for any $\xi \in E$, we have

$$\xi\langle b\eta,\ \mu\rangle_A =_B \langle \xi,\ b\eta\rangle\mu$$

$$=_B \langle \xi,\ \eta\rangle b^*\mu = \xi\langle \eta,\ b^*\mu\rangle_A.$$

Hence

$$\langle b\eta,\ \mu\rangle_A = \langle \eta,\ b^*\mu\rangle_A$$

by the above claim. Similarly,

$$_B\langle \eta a,\ \mu\rangle =_B \langle \eta,\ \mu a^*\rangle.$$

The proof of Theorem 2.2. (1) $\Rightarrow$ (2) : Suppose that $A \sim_M B$. By Definition 2.2 there exists an equivalence A-B-bimodule E which is full with respect to A-valued inner product and B-valued inner product, respectively. For any fixed $b \in B$, define a A-linear mapping by $\lambda_b(\xi) = b\xi$ $(\xi \in E)$. Since

$$||\lambda_b(\xi)|| = ||b\xi|| \leq ||b||||\xi||,$$

it implies that $||\lambda_b|| \leq ||b||$, which shows that λ_b is bounded module mapping. For any $\eta \in E$,

$$\langle \lambda_b(\xi),\ \eta\rangle_A = \langle b\xi,\ \eta\rangle_A = \langle \xi, b^*\eta\rangle_A,$$

from Remark 2.3, it follows that $(\lambda_b)^* = \lambda_{b^*}$. Thus $\lambda_b \in L_A(E)$. For element of the form $_B\langle \eta, \tau\rangle$, since

$$\lambda_{_B\langle \eta,\ \tau\rangle}(\xi) =_B \langle \eta, \tau\rangle\xi$$

$$= \eta < \tau,\ \xi >_A = \theta_{\eta,\ \tau}(\xi),$$

$\lambda_{_B\langle \eta,\ \tau\rangle} \in K_A(E)$. From the above, it implies that $\lambda_b \in K_A(E)$ $(b \in B)$ since the set of all finite linear combination of elements of form $_B\langle \eta, \tau\rangle$ is dense in B. Define a linear mapping Φ from B to $K_A(E)$ given by

$$\Phi : B \to K_A(E), b \to \lambda_b\ (b \in B).$$

Since for all b_1 and b_2 in B,, we have

$$\Phi(b_1 b_2) = \lambda_{b_1 b_2} = \Phi(b_1)\Phi(b_2),$$

and

$$\Phi(b^*) = \lambda_{b^*} = (\lambda_b)^* = (\Phi(b))^*.$$

Hence Φ is a $*$ homomorphism. Now, to show that Φ is injective. If $\Phi(b) = \lambda_b = 0$,, then $\lambda_b \xi = b\xi = 0$ for all $\xi \in E$. Since there exists an approximation unit of form $\{u_\alpha\}_{\alpha \in \Gamma}$ in B by Lemma 2.2, where $u_\alpha = \sum_{i=1}^{n(\alpha)} {}_B\langle \eta_i^{(\alpha)}, \eta_i^{(\alpha)} \rangle$, it follows that

$$b\, u_\alpha = b \sum_{i=1}^{n(\alpha)} {}_B\langle \eta_i^{(\alpha)},\ \eta_i^{(\alpha)} \rangle$$

$$= \sum_{i=1}^{n(\alpha)} {}_B\langle b\eta_i^{(\alpha)},\ \eta_i^{(\alpha)} \rangle = 0\ (\alpha \in \Gamma).$$

Therefore from the above proof that $b = 0$, which shows that Φ is an isometric mapping by using Theorem 3.1.5 in [17], the rest is to prove that Φ is surjective. Equivalently, it suffices to prove that the image of a dense subset is dense. This can be proved from the following equality $\Phi(\sum_{i=1}^{n} {}_B\langle \xi_i, \eta_i \rangle) = \sum_{i=1}^{n} \theta_{\xi_i, \eta_i}$ for all $\xi_i,\ \eta_i \in E,\ \ i = 1, 2, \cdots, n$, since $\overline{{}_B\langle E,\ E \rangle} = B$ and the set of all elements of the form $\sum_{i=1}^{n} \theta_{\xi_i,\ \eta_i}$ is dense in $K_A(E)$.

(2) $\Rightarrow$ (1) : Suppose that $\Phi : B \to K_A(E)$ is an isomorphic surjection. For any $b \in B,\ \xi \in E$, let $b\, \xi = \Phi(b)\xi$, so that E becomes a left B-module. Then define a B-value inner product given by

$${}_B\langle \xi, \eta \rangle = \Phi^{-1}(\theta_{\xi,\eta})\ (\xi, \eta \in E).$$

It follows that $\overline{{}_B\langle E, E \rangle} = \Phi^{-1}(K_A(E)) = B$, which shows that E is full with B-value inner product, since the set of all finite linear combinations of elements of the form $\theta_{\xi,\eta}$ is dense in $K_A(E)$. Continue the above process, we have for any $\xi,\ \eta,\ \tau \in E$,

$${}_B\langle \xi, \eta \rangle \tau = \Phi({}_B\langle \xi,\ \eta \rangle)\tau$$

$$= \Phi(\Phi^{-1}(\theta_{\xi,\ \eta}))\tau = \theta_{\xi,\ \eta}(\tau).$$

Since $\xi \langle \eta,\ \tau \rangle_A = \theta_{\xi,\ \eta}(\tau)$, which follows that $\xi \langle \eta,\ \tau \rangle_A = {}_B\langle \xi,\ \eta \rangle \tau$. Thus $A \sim_M B$ by Definition 2.2.

Remark 2.4

(1) In many literatures, the condition (2) in Theorem 2.2 is directly used as the definition of strong Morita equivalence.

(2) Definition 2.2 generalizes the concept of M.A.Rieffel's imprimitivity bimodule (see Definition 6.10 in [20]). The imprimitivity bimodule E not only requires that it is full with respect to A-valued inner product and B-valued inner product respectively but also satisfies the following conditions:

1° ${}_B\langle \xi a,\ \xi a \rangle \leq ||a||^2 {}_B\langle \xi,\ \xi \rangle$,

and

2° $\langle b\xi,\ b\xi \rangle_A \leq ||b||^2 \langle \xi,\ \xi \rangle$,

for all $a \in A,\ \ b \in B,\ \ \xi \in E$.

In fact, the above condition 1° and 2° can be simply deduced from Definition 2.2. Since ${}_B\langle \xi a,\ \xi a\rangle = {}_B\langle \xi,\ \xi aa^*\rangle$ from Remark 2.3, and since $aa^* \le ||a||^2 1$, where 1 is the unit of $\widetilde{A} = A \oplus \mathbb{C}$, it follows that

$$ {}_B\langle \xi,\ \xi(||a||^2 1 - aa^*)\rangle $$

$$ = {}_B\langle \xi(||a||^2 1 - aa^*)^{1/2},\ \xi(||a||^2 1 - aa^*)^{1/2}\rangle \ge 0. $$

Therefore the condition 1° is proved. Similarly, the condition 2° is also true.

(3) If E ia an equivalence A-B-bimodule, then

$$ ||{}_B\langle \xi,\ \xi\rangle|| = ||\langle \xi,\ \xi\rangle_A|| \ (\xi \in E). $$

It is shown that the norm induced by A-valued inner product is the same as that induced by B-valued inner product. In fact, from Proposition 1.4, we see that

$$ ||\theta_{\xi,\ \xi}|| = ||<\xi,\ \xi>_A|| \ (\xi \in E). $$

Since ${}_B\langle \xi,\ \xi\rangle \to \lambda_{{}_B\langle \xi,\xi\rangle} = \theta_{\xi,\ \xi}$, $||{}_B\langle \xi, \xi\rangle|| = ||\theta_{\xi,\ \xi}||$ because $b \to \lambda_b \in L_A(E)$ $(b \in B)$ is isometric by Theorem 2.2. Thus $||_B < \xi,\ \xi > || = || < \xi,\ \xi >_A ||$.

Example 2.1 Suppose that A is a C^*-algebra and E is a Hilbert A-module. From $K(E)E \subset E$, we see that E is a left $K(E)$-module. Then E becomes an equivalence A-$K(E)$-bimodule when endowed with the following $K(E)$-valued inner product:

$$ {}_{K(E)}\langle \xi, \eta\rangle = \theta_{\xi,\eta} \ (\xi,\ \eta \in E). $$

Moreover, if E is full as A-module, then it is easy to check that $K(E) \sim_M A$.

Example 2.2 Let D be a C^*-algebra, and let A and B be its C^*-subalgebra. Suppose that closed subspace L in D satisfies the following conditions:

$$ BL \subset L;\ LA \subset L;\ LL^* \subset B;\ L^*L \subset A. $$

Then we can define A-valued inner product and B-valued inner product on E given by

$$ \langle r,\ s\rangle_A = r^* s, $$

and

$$ {}_B\langle r, s\rangle = rs^*, $$

for all $r,\ s \in L$. Since

$$r\langle s,\ t\rangle_A = rs^*t$$
$$={}_B\langle r,\ s\rangle t,$$

L is an equivalence A-B-bimodule.

The following property shows that Morita equivalence is indeed an equivalence relation.

Proposition 2.5 *Morita equivalence is an equivalence relation, that is*

(1) $A \sim_M A$;
(2) If $A \sim_M B$ *then* $B \sim_M A$;
(3) If $A \sim_M B$ *and* $B \sim_M C$, *then* $A \sim_M C$.

Proof

(1) Since A is full when A itself is regarded as Hilbert A-module, and $A \cong K_A(A)$ from Example 1.5. Therefore $A \sim_M A$ by Theorem 2.2.
(2) Suppose that $A \sim_M B$. Then there exists a full Hilbert A-module E such that $B \cong K_A(E)$ by Theorem 2.2. In what follows, without loss of generality, we assume that $B = K_A(E)$. Let $F = K_A(E,\ A)$. Then F is a right B-module since $K_A(E, A)K_A(E) \subset K_A(E,\ A)$. Since

$$a_1\theta_{a_2,\ \xi} = \theta_{a_1a_2,\ \xi} \in K_A(E,\ A),$$

for all $a_1,\ a_2 \in A$ and $\xi \in E$. Then $AK_A(E,\ A) \subset K_A(E,\ A)$. Thus F is also a left A-module. Then define A-valued and B-valued inner products on F by the following way:

$${}_A\langle\theta_{a,\ \xi},\ \theta_{a',\ \eta}\rangle = \theta_{a,\ \xi}\cdot\theta^*_{a',\ \eta} \in K(A) = A,$$

and

$$\langle\theta_{a,\ \xi}, \theta_{a',\ \eta}\rangle_B = \theta^*_{a,\ \xi}\cdot\theta_{a',\ \eta} \in B,$$

for all $\xi,\ \eta \in E$ and $a,\ a' \in A$. Since $\overline{E < E,\ E >} = E$ by Proposition 1.1, the set of all finite linear combination of elements of form $\theta_{\xi a^* a',\ \eta}$ is dense in $K_A(E)$. And since

$$\theta^*_{a,\ \xi}\cdot\theta_{a',\ \eta} = \theta_{\xi,\ a}\theta_{a',\ \eta}$$
$$= \theta_{\xi <a,\ a'>,\eta} = \theta_{\xi a^* a',\ \eta}.$$

It follows that $\overline{< F, F >_B} = B$,; this shows that F is full with respect to B-valued inner product. Similarly, F is also full with A-valued inner product. When $\theta_{a,\xi}, \theta_{b,\eta}, \theta_{c,\tau} \in F, \xi, \eta, \tau \in E$ and $a, b, c \in A$, since

$$\theta_{a,\xi}\langle\theta_{b,\eta},\theta_{c,\tau}\rangle_B$$

$$= \theta_{a,\xi} \cdot \theta_{b,\eta}^{*}\theta_{c,\tau}$$

$$=_A \langle\theta_{a,\xi}, \theta_{b,\eta}\rangle\theta_{c,\tau}.$$

and since the set of all finite linear combination of elements of form $\theta_{a,\xi}$ is dense in F, it results from the linearity and continuity of inner product that

$$_A\langle x, y\rangle z = x\langle y, z\rangle_B,$$

for all $x, y, z \in F$. Therefore $B \sim_M A$ by Definition 2.2.

(3) Suppose that $A \sim_M B$ and $B \sim_M C$. From Theorem 2.2, there exists a full Hilbert A-module E such that $B \cong K_A(E)$, which is denoted by $\varphi : B \to K_A(E)$. Meanwhile, there exists a full Hilbert B-module F such that $C \cong K_B(F)$. Set interior tensor product $F \otimes_\varphi E$ of E and F, then $F \otimes_\varphi E$ is a full Hilbert A-module when endowed with the following A-valued inner product:

$$< y_1 \otimes_\varphi x_1, y_2 \otimes_\varphi x_2 >_A =< x_1, \varphi(< y_1, y_2 >)x_2 >_A$$
$$(y_1 \otimes_\varphi x_1, y_2 \otimes_\varphi x_2 \in F \otimes_\varphi E),$$

since $K_A(E)E = E$ and φ is $*$ isomorphism. In order to show $A \sim_M C$, we need to prove that $C \cong K_A(F \otimes_\varphi E)$ by Theorem 2.2. It is equivalent to prove that $K_B(F) \cong K_A(F \otimes_\varphi E)$, since $C \cong K_B(F)$. For any fixed $x, y, u \in F$ and $v \in E$, define a A-linear mapping given by

$$\varphi_*(\theta_{x,y}) : F \otimes_\varphi E \to F \otimes_\varphi E, u \otimes_\varphi v \to \theta_{x,y}(u) \otimes_\varphi v.$$

By routine calculation, $\varphi_*(\theta_{x,y})$ can be extended to an element in $L_A(F \otimes_\varphi E)$, which is still denoted as $\varphi_*(\theta_{x,y})$, and its adjoint $(\varphi_*(\theta_{x,y}))^* = \varphi_*(\theta_{y,x})$. Furthermore, it is easy to check that the linear mapping $\varphi_* : K_B(F) \to L_A(F \otimes_\varphi E)$ given by

$$\varphi_*(\theta_{x,y})(u \otimes_\varphi v) = \theta_{x,y}(u) \otimes_\varphi v$$

is a $*$ homomorphism.

In the following, to see that φ_* is injective. For this, it suffices to prove that if $\varphi_*(\theta_{x,y}) = 0$ then $\theta_{x,y} = 0$, since the set of all finite linear combination of elements of the form $\theta_{x,\ y}$ is dense in $K_B(F)$. Indeed, if for any $u \otimes_\varphi v \in F \otimes_\varphi E$,, we have $\varphi_*(\theta_{x,\ y})(u \otimes_\varphi v) = 0$, that is, $\theta_{x,\ y}(u) \otimes_\varphi v = 0$. It follows that

$$< \theta_{x,\ y}(u) \otimes_\varphi v,\ u' \otimes_\varphi v' >$$

$$=< v, \varphi(< \theta_{x,\ y}u,\ u' >)v' >_A = 0,$$

for all $u' \otimes_\varphi v' \in F \otimes_\varphi E$. Because v and v' are arbitrary in E, so $\varphi(< \theta_{x,\ y}u,\ u' >) = 0$, from which it shows that $< \theta_{x,\ y}u,\ u' >= 0$ since φ is injective. Thus $\theta_{x,\ y} = 0$ because u and u' are arbitrary in F.

Finally, we have to prove that $\varphi_*(K_B(F)) = K_A(F \otimes_\varphi E)$. When $z,\ w \in E$, there exists an element $b \in B$ such that $\varphi(b) = \theta_{z,\ w}$ since φ is surjective. Therefore

$$\theta_{x\otimes_\varphi z,\ y\otimes_\varphi w}(u \otimes_\varphi v)$$

$$= x \otimes_\varphi z < y \otimes_\varphi w,\ u \otimes_\varphi v >$$

$$= x \otimes_\varphi z < w,\ \varphi(< y,\ u >)v >$$

$$= x \otimes_\varphi (\theta_{z,\ w}\varphi(< y,\ u >)v)$$

$$= x \otimes_\varphi \varphi(b < y,\ u >)v$$

$$= xb < y, u > \otimes_\varphi v$$

$$= \theta_{xb,\ y}(u) \otimes_\varphi v$$

$$= \varphi_*(\theta_{xb,\ y})u \otimes_\varphi v.$$

The above proof shows that $\varphi_*(K_B(F)) \supset K_A(F \otimes_\varphi E)$ since $K_A(F \otimes_\varphi E)$ is generated linearly by elements of the form $\theta_{x\otimes_\varphi z,\ y\otimes_\varphi w}$.

Because $\varphi(b) \in K_A(E)$ for all $b \in B$, so there exists a subsequence $(n_k)_{k=1}^\infty$ in $\mathbb{N}$ and corresponding sequences $x_i^{(n_k)}$ and $y_i^{(n_k)}$ in E $(i = 1,\ 2,\ \cdots,\ n_k,\ k \in N)$, such that

$$\lim_{k\to+\infty} || \sum_{i=1}^{n_k} \theta_{x_i^{(n_k)},\ y_i^{(n_k)}} - \varphi(b)|| = 0.$$

Hence

$$\varphi_*(\theta_{yb,\ x})u \otimes_\varphi v = \theta_{yb,\ x}(u) \otimes_\varphi v$$

$$= yb < x,\ u > \otimes_\varphi v = y \otimes_\varphi \varphi(b < x,\ u >)v$$

$$= y \otimes_\varphi \varphi(b)\varphi(< x,\ u >)v$$

$$= \lim_{k \to +\infty} \sum_{i=1}^{n_k} y \otimes_\varphi \theta_{x_i^{(n_k)},\ y_i^{(n_k)}} \varphi(< x,\ u >)v$$

$$= \lim_{k \to +\infty} \sum_{i=1}^{n_k} y \otimes_\varphi x_i^{(n_k)} < y_i^{(n_k)},\ \varphi(< x,\ u >)v >$$

$$= \lim_{k \to +\infty} \sum_{i=1}^{n_k} y \otimes_\varphi x_i^{(n_k)} < x \otimes_\varphi y_i^{(n_k)},\ u \otimes_\varphi v >,$$

$$= \lim_{k \to +\infty} \sum_{i=1}^{n_k} \theta_{y \otimes_\varphi x_i^{(n_k)},\ x \otimes_\varphi y_i^{(n_k)}} (u \otimes_\varphi v),$$

for all $x,\ y \in F$ and $u \otimes_\varphi v \in F \otimes_\varphi E$. Therefore

$$\varphi_*(\theta_{yb,\ x}) = \lim_{k \to +\infty} \sum_{i=1}^{n_k} \theta_{y \otimes_\varphi x_i^{(n_k)},\ x \otimes_\varphi y_i^{(n_k)}} \in K_A(F \otimes_\varphi E).$$

It implies that $\varphi_*(K_B(F)) \subset K_A(F \otimes_\varphi E)$, since $\overline{F < F,\ F >} = F$ and $K_B(F)$ is generated linearly by elements of the form $\theta_{yb,\ x}$. Hence $\varphi_*(K_B(F)) = K_A(F \otimes_\varphi E)$, which shows that $K_B(F) \cong K_A(F \otimes_\varphi E)$.

In what follows, we shall describe the relationship between stable isomorphism and Morita equivalence of C^*-algebras. It is easy to check that if C^*-algebra A and C^*-algebra B are $*$ isomorphic then $A \sim_M B$. In fact, since A is full when it is self as a Hilbert A-module, we have $K(A) \cong A \cong B$, from which it shows that $A \sim_M B$ by Theorem 2.2. Since

$$K(H \otimes A) \cong K(H) \otimes K(A) \cong K(H) \otimes A,$$

for all C^*-algebra A. Note that $H \otimes A$ is full as Hilbert A-module, hence $A \sim_M K(H) \otimes A$ by Theorem 2.2. Therefore if A and B are stable isomorphic, namely, if $K(H) \otimes A \cong K(H) \otimes B$, then by the above observations,

$$A \sim_M K(H) \otimes A \sim_M K(H) \otimes B \sim_M B.$$

Thus $A \sim_M B$ by Proposition 2.5.

The above conclusion is also invertible if proper separability conditions are added; this is the following stable isomorphism theorem of L.Brown, P.Green, and M.Rieffel (see [4]).

Theorem 2.3 *Two σ-unital C^*-algebras A and B are stably isomorphic if and only if they are Morita equivalent, namely, $A \sim_M B$.*

We need a Lemma.

Lemma 2.3 *Let E be a full Hilbert C^*-module over a σ-unital C^*-algebra A, and let H be a separable Hilbert space. Then the following conclusions hold;*

(1) There exists a Hilbert A-module F such that $H \otimes E \cong A \oplus F$;
(2) If E is countably generated, then $H \otimes E \cong H \otimes A$.

Proof

(1) Since A is σ-unital and E is full, from Corollary 2.2, there exists sequence $(x_n) \subset E$ such that $\{\sum_{i=1}^{n} < x_i, x_i >\}_{n=1}^{\infty}$ is an approximate unit element of A. Hence for any $a, b \in A$,, we have

$$\sum_{n=1}^{\infty} a < x_n, x_n > b = ab,$$

in the sense of norm convergence on A. Write $\{\varepsilon_n\}$ as a standard orthogonal basis of H. Define a A-linear mapping from A to $H \otimes E$ given by

$$T(a) = \sum_{n=1}^{\infty} \varepsilon_n \otimes x_n a \ (a \in A).$$

Since

$$||T(a)||^2 = || < T(a), T(a) > ||$$

$$= ||\sum_{n=1}^{\infty} \varepsilon_n \otimes x_n, \sum_{m=1}^{\infty} \varepsilon_m \otimes x_m||$$

$$= ||\sum_{m=1}^{\infty}\sum_{n=1}^{\infty} < \varepsilon_n, \varepsilon_m >< x_n a, x_m a > ||$$

$$= \sum_{n=1}^{\infty} a^* < x_n, x_n > a = ||a^* a|| = ||a||^2.$$

Hence T is an isometric. Since

$$< T(a),\ \varepsilon_n \otimes x >_A$$

$$=< \sum_{m=1}^{\infty} \varepsilon_m \otimes x_m a,\ \varepsilon_n \otimes x >$$

$$=< x_n a,\ x >= a^* < x_n,\ x >$$

$$=< a,\ < x_n,\ x >>_A,$$

from which it shows that $T^*(\varepsilon_n \otimes x) =< x_n,\ x >$. Thus

$$T^*T(a) = T^*(\sum_{n=1}^{\infty} \varepsilon_n \otimes x_n a)$$

$$= \sum_{n=1}^{\infty} T^*(\varepsilon_n \otimes x_n a) = \sum_{n=1}^{\infty} < x_n,\ x_n a >$$

$$= \sum_{n=1}^{\infty} < x_n,\ x_n > a = a.$$

So $T^*T = I$ is the identity module mapping. Therefore TT^* is a projection in $L(H \otimes E)$, and it is easy to check that $R(TT^*) = R(T)$, which shows that $R(T)$ is a complemented submodule in $H \otimes E$. Let $F = R(T)^{\perp}$, then $H \otimes E = R(T) \oplus F$. From the above proof, we see that $T : A \to R(T)$ is an isometric surjection. Thus $A \cong R(T)$ by Theorem 1.2. Therefore $H \otimes E \cong A \oplus F$.

(2) Note that $H \cong H \otimes H$, then from (1), we have

$$H \otimes E \cong H \otimes H \otimes E$$

$$\cong H \otimes (A \oplus F) \cong (H \otimes A) \oplus (H \otimes F).$$

Note that E is countably generated, so is $H \otimes E$ also. Hence, $F(\subset H \otimes E)$ and $H \otimes F$ are countably generated, respectively. Therefore

$$(H \otimes A) \oplus (H \otimes F) \cong H \otimes A,$$

by Kasprove's stabilization theorem (see Theorem 2.1). Thus $H \otimes E \cong H \otimes A$ from the above proof.

The proof of Theorem 2.3. The necessity has been proved above. We just need to prove the sufficiency. Suppose that $A \sim_M B$. Then from Theorem 2.2, there exists a full Hilbert A-module E such that $B \cong K_A(E)$. Since B is σ-*unital*, $K_A(E)$ is σ-*unital* also. Hence E is countably generated by Proposition 2.4. Thus, combining Lemma 2.3, we get

$$K(H) \otimes B \cong K(H) \otimes K_A(E)$$

$$\cong K_A(H \otimes E) \cong K_A(H \otimes A)$$

$$\cong K(H) \otimes K(A) \cong K(H) \otimes A.$$

2.2.2 C^*-Correspondences and Cuntz-Pimsner Algebras

Cuntz-Pimsner algebra was introduced by M.V.Pimsner in [19], which is based on a special bimodule, that is C^*-correspondence, which is an important concept in noncommutative geometry theory [2]. Cuntz-Pimsner algebra has important applications in operator valued free probability theory [22].

Cuntz-Pimsner algebra generalizes the Cuntz algebra. Let us first briefly describe the Cuntz algebra. Given n $(2 \leq n \leq \infty)$ isometric operators $\{s_i\}_{i=1}^n$ which satisfy the following conditions:

$$\sum_{i=1}^{\infty} s_i s_i^* \leq I, \ if \ n = \infty;$$

$$\sum_{i=1}^{n} s_i s_i^* = I, \ if \ 2 \leq n < \infty,$$

where I is the identity operator. Then the C^*-algebra generated by $\{s_i\}_{i=1}^n$ is called Cuntz algebra, which is denoted as O_n. In particular, when $n = \infty$, it is recorded as O_∞. J. Cuntz [6] proved that if $\{s_i\}_{i=1}^n$ and $\{t_i\}_{i=1}^n$ are isometric operators families satisfying the above conditions, then the C^*-algebras generated by them, respectively, are isomorphic, and O_n $(n \in \mathbb{N})$ are all simple C^*-algebras. Furthermore, in [6], the following conclusion is obtained:

Proposition 2.6 (see [6], Proposition 3.1) *Let $v_1, v_2, \cdots, v_n (2 \leq n < \infty)$ be n isometric operators and $\sum_{i=1}^n v_i v_i^* \leq I$. Denote by $C^*(v_1, \cdots, v_n)$ the C^*-algebra generated by $\{v_i\}_{i=1}^n$. Then the closed two-sided ideal J of $C^*(v_1, \cdots, v_n)$ generated by projection operator $I - \sum_{i=1}^n v_i v_i^*$ has the properties that J is $*$ isomorphic to the compact operator ideal $K(H)$ for a separable Hilbert space H, and quotient algebra $C^*(v_1, \cdots, v_n)/J \cong O_n$.*

Let us look at Cuntz algebras from the perspective of Fock spaces. Given a n-dimension Hilbert space H, $\{e_i\}_{i=1}^n$ is a standard orthogonal basis of H. Construct

Fock space by the following way:

$$F(H) = \mathbb{C} \oplus H \oplus H^{\otimes 2} \oplus \cdots = \sum_{k=0}^{\infty} \bigoplus H^{\otimes k},$$

where $H^{\otimes 0} = \mathbb{C}$. Define isometric operators $v_i \in B(F(H))$ $(i = 1,\ 2,\ \cdots,\ n)$ given by

$$v_i x = e_i \otimes x, \forall x \in H^{\otimes k},\ i = 1,\ 2,\ \cdots,\ n;\ k = 0,\ 1,\ 2,\ \cdots.$$

From the above, it is easy to check that the adjoint v_i^* of v_i is given by

$$v_i^* x = 0,\ \text{if } x \in \mathbb{C};$$

$$v_i^* x_1 \otimes x_2 \otimes \cdots x_k =< e_i,\ x_1 > x_2 \otimes \cdots x_k,\ \text{if } k \geq 1,$$

and $\sum_{i=1}^{n} v_i v_i^* = I - P_0$, where P_0 is the projection from $F(H)$ to $\mathbb{C}$ such that $P_0(x_0 + x_1 + \cdots +) = x_0$.

Furthermore, one can obtain that $O_n \cong C^*(v_1, \cdots v_n)/J$ when $n < \infty$, where $J = K(F(H))$ is the closed two-sided ideal of $C^*(v_1, \cdots, v_n)$ generated by P_0; $O_\infty \cong C^*(v_1, \cdots, v_n)$ when $n = \infty$.

The starting point of Cuntz-Pimsner algebra is to change the above Hilbert space into a special class of bimodules, namely, C^*-correspondence. The concept of C^*-correspondence is given as follows:

Definition 2.3 A Hilbert A-module E is said to be a C^*-correspondence over C^*-algebra A provided that there exists a $*$-homomorphism $\phi : A \to L(E)$ such that E is a A-A-bimodule when endowed with the left action of A on E given by

$$a \cdot x = \phi(a)x\ (a \in A, x \in E).$$

A C^*-correspondence E over a C^*-algebra A is said to be faithful if and only if map ϕ is faithful. It is called strict if $\overline{[\phi(A)E]}$ is complemented in E. In particular, if $\overline{[\phi(A)E]} = E$, i.e., the map ϕ is nondegenerate, then E is said to be essential. If F is also a C^*-correspondence over A, so is still the interior tensor product $E \otimes_\phi F$.

The C^*-correspondences we consider in the following are faithful and essential. Since A has approximate unit, then $E = \phi(A)E$ by Cohen factorization theorem.

Definition 2.4 Let E and F be C^*-correspondences over A and B, respectively. A covariant homomorphism from (A, E) to (B, F) is a pair $(\sigma,\ \phi)$, where $\sigma : A \to B$ is a $*$ homomorphism and $\phi : E \to F$ is a linear mapping such that for all $\xi,\ \eta \in E$ and $a_1,\ a_2 \in A$,

$$\phi(a_1 \xi a_2) = \sigma(a_1)\phi(\xi)\sigma(a_2),$$

and

$$< \phi(\xi), \phi(\eta) >= \sigma(< \xi, \eta >).$$

Moreover, if ϕ is bijection, then $(\sigma,\ \phi)$ is called an isomorphism from E to F, denoted by $E \cong F$.

In the following, the concept of Morita equivalence between C^*-correspondences is given.

Definition 2.5 Let E and F be C^*-correspondence over C^*-algebras A and B, respectively. If there exists an equivalence A-B -bimodule X and correspondence isomorphism mapping $\phi : X \otimes_B F \to E \otimes_A X$, then E and F are called Morita equivalence, denoted by $E \sim_M F$.

Similar to the case of Fock space on Hilbert space, the Hilbert A-module

$$F(E) = A \oplus E \oplus E^{\otimes 2} \oplus \cdots \oplus E^{\otimes n} \oplus = \sum_{k=0}^{\infty} \bigoplus E^{\otimes k},$$

is called full Fock module space on E, where $E^{\otimes 0} = A$, and each term $E^{\otimes n} (n \geq 2)$ is the interior tensor product of n times of E.

For example, when $n = 2$, the A-valued inner product is given by

$$< \xi_1 \otimes \eta_1,\ \xi_2 \otimes \eta_2 >=< \eta_1,\ \phi(< \xi_1,\ \xi_2 >)\eta_2 > .$$

The right A-module action and the left A-module action are given, respectively, by

$$(\xi_1 \otimes \xi_2)a = \xi_1 \otimes (\xi_2 a),$$

and

$$\phi_2(a)(\xi_1 \otimes \xi_2) = (\phi(a)\xi_1) \otimes \xi_2.$$

Similarly, $\phi_3,\ \phi_4,\ \cdots,\ \phi_n,\ \cdots$, can be defined.

$\phi_\infty(a) \in L(F(E))$ is defined by

$$\phi_\infty(a)(\xi_1 \otimes \cdots \otimes \xi_k) = \phi(a)\xi_1 \otimes \cdots \otimes \xi_k\ (a \in A,\ k \in \mathbb{N}).$$

Suppose that ξ is in E. Then the creation operator T_ξ on $F(E)$ determined by ξ is defined by the following way

$$T_\xi \eta = \begin{cases} \xi \otimes \eta, \text{ if } \eta \in E^{\otimes k},\ k > 0; \\ \xi a, \quad \text{if } \eta = a \in E^{\otimes 0} = A. \end{cases}$$

It is easy to check that T_ξ has adjoint given by

$$T_\xi^*(\eta_1 \otimes \eta) = \phi_k(<\xi,\ \eta_1>)\eta\ (\eta_1 \in E,\ \eta \in E^{\otimes k},\ k \in \mathbb{N}),$$

which is called annihilation operator. Thus $T_\xi \in L(F(E))$.

The Toeplitz C^*-algebra associated with the correspondence E is the C^*-subalgebra in $L(F(E))$ generated by $\{T_\xi : \xi \in E\} \cup \{\phi_\infty(a) : a \in A\}$, denoted by $T(E)$. In fact, if $E = A = \mathbb{C}$, ϕ is the identity, then $F(E)$ is $l^2(\mathbb{C})$ and $T(E)$ is the Toeplitz C^*-algebra generated by one-sided shift operators, respectively. If $E = H$ is n-dimension Hilbert space, then $T(E)$ is the previously defined C^*-algebra $C^*(v_1, \cdots, v_n)$.

Definition 2.6 Quotient algebra $T(E)/T(E) \cap K(F(E))$ is called Cuntz-Pimsner algebra, which is denoted by O_E.

Remark 2.5

(1) If $E \sim_M F$ then $O_E \sim_M O_F$. (see [16])
(2) If we write $J = \phi^{-1}(K(E)) = \{a \in A : \phi(a) \in K(E)\}$, then J is a closed two-sided ideal of A (including the trivial case $J = 0$, or A). From Theorem 3.13 in [19], we see that $T(E) \cap K(F(E)) = K(F(E)J)$. If E is a finitely generated projective module, then $O_E = T(E)/K(F(E))$ since $K(F(E)) \subset F(E)$. In particular, if E is a n-dimensional Hilbert space, then $O_E = O_n$.

In generally, O_E is not simple C^*-algebra. Therefore, under what conditions O_E is a simple C^*-algebra which has once become a hot topic in the research of Cuntz-Pimsner algebra. The literature in this field can be referred to [12, 16], etc.

2.3 Generalized Index Theory of Fredholm Module Operators

2.3.1 A Brief Introduction to Fredholm Operators Theory on Hilbert Spaces

First it is emphasized here that for the convenience of description, all C^*-algebras in this section have units unless otherwise specified.

Suppose that H is an infinite dimensional Hilbert space. An operator T in $B(H)$ is called a finite rank operator if the dimension of the range of T is finite and a compact operator if the image of the closed unit ball of H under of T is a compact subset of H, respectively. Let $F(H)$ denote the set of all finite rank operators and $K(H)$ the set of all compact operators.

It is well known that $F(H)$ is a minimal two-sided ideal in $B(H)$, and $F(H)$ is dense in $K(H)$ under the norm topology. If H is separable, then $K(H)$ is the only proper closed two-sided ideal in $B(H)$. Therefore, H is finite dimensional if and only if $K(H) = B(H)$, equivalently, the closed unit ball of H is a compact subset in norm topology.

The following proposition gives an equivalent characterization of compact operators.

Proposition 2.7 (see [9], Corollary 5.10) *If H is an infinite dimensional Hilbert space and operator T in $B(H)$. Then T is compact if and only if the range of T contains no closed infinite dimensional subspaces.*

Now we introduce briefly the index theory of Fredholm operators which is an important part of operator theory.

Definition 2.7 Let H be a Hilbert space, and let T be in $B(H)$. Then T is called a Fredholm operator if its range $R(T)$ is closed in H and if the dimensions of $kerT$ (write as $dimkerT$) and the dimension of $kerT^*$ (write as $dimkerT^*$) are finite, respectively.

The set of all Fredholm operators is denoted by $\Phi(H)$.

Remark 2.6 Since $kerT^* = R(T)^{\perp}$, $dimkerT^*$ is called the codimension of $R(T)$.

If T is invertible in $B(H)$, then T is Fredholm operator by Definition 2.7 since $kerT = kerT^* = \{0\}$. This shows that every invertible operator in $B(H)$ is Fredholm operator. In particular, if H is a finite dimensional space, then every operator in $B(H)$ is Fredholm operator, in this case, $\Phi(H) = B(H)$. Therefore, we assume that the Hilbert spaces appearing in the remainder of this chapter are all infinite dimensional.

Atkinson gave an equivalent characterization of Fredholm operators by using Calkin algebra, that is, quotient algebra $B(H)/K(H)$, which is denoted as $C(H)$. The natural homomorphism $\pi : B(H) \to B(H)/K(H)$ is a continuous open and surjective mapping.

Proposition 2.8 (Atkinson theorem) *If H is a Hilbert space and operator T is in $B(H)$. Then T is a Fredholm operator if and only if $\pi(T)$ is an invertible element in $C(H)$.*

Proof If T is a Fredholm operator, then $R(T)$ is a closed subspace in H, $kerT$, and $kerT^*$ are finite dimensional subspaces in H by Definition 2.7. When T is restricted to $(kerT)^{\perp}$, then $T|_{(kerT)^{\perp}} : (kerT)^{\perp} \to R(T)$ is a bijection. Hence $T_{(kerT)^{\perp}}$ is invertible by the Banach inverse operator theorem. Now we define a bounded linear operator $S : H \to H$ given by

$$Sx = \begin{cases} T^{-1}_{(kerT)^{\perp}}x, & \text{if } x \in R(T); \\ 0, & \text{if } x \in R(T)^{\perp}. \end{cases}$$

From the construction of S, we have $ST = I - P_1$ and $TS = I - P_2$, where I is the unit in $B(H)$ and P_1 is the projection from H onto $kerT$, P_2 is the projection from H onto $R(T)^{\perp} = kerT^*$. It follows that

$$\pi(S)\pi(T) = \pi(T)\pi(S) = \pi(I),$$

which shows that $\pi(S)^{-1} = \pi(T)$.

Conversely, if $\pi(T)$ is an invertible element in $C(H)$, then there exist compacts K_1, K_2 and operators A_1 and A_2 in $B(H)$ such that $A_1T = I + K_1$ and $TA_2 = I + K_2$. Hence

$$kerT \subset kerA_1T = ker(I + K_1),$$

and

$$kerT^* \subset ker(A_2^*T^*) = ker(I + K_2^*).$$

To show $kerT$ is finite dimensional, it suffices to prove that $ker(I + K_1)$ is finite dimensional. Write $Y = ker(I + K_1)$. Since $(I + K_1)y = 0$ for all $y \in Y$, $K_1y = -y \in Y$, from which it shows that $K_1Y \subset Y$ and K_1 equals $-I_Y$ when restricted to Y, where I_Y is identity operator on Y. From the previous basic conclusion about compact operators, we see that Y is a finite dimensional subspace. Similarly, it can be proved that $ker(I + K_2^*)$ is also.

Finally, we want to prove that $R(T)$ is closed. Choose a finite rank operator F such that $||K_1 - F|| < \frac{1}{2}$, hence

$$||A_1||||Tx|| \geq ||A_1Tx||$$

$$= ||x + K_1x|| = ||x + Fx + K_1x - Fx||$$

$$\geq ||x|| - ||K_1x - Fx|| \geq \frac{1}{2}||x||,$$

from which it implies that $||Tx|| \geq \frac{1}{2}||A_1||^{-1}||x||$,; this shows that T has lower bound when it is restricted to $kerF$. Therefore $T(kerF)$ is a closed subspace in H. Thus

$$R(T) = T(kerF + (kerF)^{\perp})$$

$$= T(kerF) + T((kerF)^{\perp})$$

is a closed subspace in H, since $(kerF)^{\perp} = R(F^*)$ is a finite dimensional subspace.

Although $dimkerT$ and $dimkerT^*$ are important for characterizing the properties of T, their difference $dimkerT - dimkerT^*$ is particularly important, since it is invariant under the small perturbation of T, which is called the Fredholm index. The details are as follows.

Definition 2.8 Let H be a Hilbert space and an operator T be in $\Phi(H)$. Then $dimkerT - dimkerT^*$ is called Fredholm index of T, which is denoted by Index(T).

For $n \in \mathbb{Z}$, set $\Phi_n = \{T \in \Phi(H) : Index(T) = n\}$, $n \in \mathbb{Z}$, which is the path connected branch of $\Phi(H)$. The main properties of Fredholm index are given below without proof. Referring to [5] for details.

Proposition 2.9 *$\Phi(H)$ is an open subset in $B(H)$ and the index mapping*

$$Index : \Phi(H) \to \mathbb{Z}, T \to Index(T) = dimkerT - dimkerT^*$$

is a continuous surjective homomorphism, that is,

$$Index(T_1T_2) = Index(T_1) + Index(T_2),$$

for all T_1 and T_2 in $\Phi(H)$, and the index mapping is invariant under compact perturbation, namely,

$$Index(T + K) = Index(T),$$

for all K in $K(H)$ and T in $\Psi(H)$.

According to Proposition 2.9, every connected component Φ_n of $\Phi(H)$ is not empty. In addition, the Fredholm operator with zero index has special significance. From Proposition 2.9, we see that if $T \in \Phi_0$ then $T + K \in \Phi_0$ for all K in $K(H)$. In particular, if T is an invertible operator, then $Index(T) = 0$, from which implies that $Index(T + K) = 0$ for all K in $K(H)$. Hence $T + K \in \Phi_0$. Surprisingly, the inverse proposition of the above conclusion is also true.

Proposition 2.10 *If $T \in \Phi(H)$ and $Index(T) = 0$, then there exists a compact operator K in $K(H)$ such that $T + K$ is invertible.*

Fredholm index is invariant not only under compact perturbation but also under small perturbation.

Proposition 2.11 *If $T \in \Phi(H)$, then there exists $\delta > 0$, when operator S in $B(H)$ satisfying $||S|| < \delta$, we have*

$$T + S \in \Phi(H), \text{ and } Index(T + S) = Index(T).$$

Remark 2.7 The Fredholm index theory on Hilbert spaces can be extended to the cases of Banach spaces and Banach algebras, respectively. For details, please refer to [1].

2.3.2 A Brief Approach to K_0-Group of C^*-Algebras

We shall need three types of equivalence of projections (denoted by $\sim$, $\sim_u$, $\sim_h$, respectively) as follows.

Definition 2.9 Let A be a C^*-algebra and projections p and q in A. If there exists v in A such that $p = v^*v$ and $q = vv^*$, then p and q are said to be equivalent, which is denoted by $p \sim q$. In this case, v is a partial isometric operator. If there exists unitary element u in A such that $p = u^*qu$, then p and q are called unitary equivalence, denoted by $p \sim_u q$. If there exists continuous mapping $\phi : [0, 1] \to P(A)$ such that $\phi(0) = p$ and $\phi(1) = q$, then p and q are called homotopy, denoted by $p \sim_h q$, where $P(A)$ is the set of all projections in A with norm topology induced by A.

Remark 2.8

(1) The above three relations satisfy reflexivity, symmetry, and transitivity. So they are all equivalent relations, and homotopy is stronger than the unitary equivalence whose stronger than the relation of equivalence. However, we shall see later when working with stable algebra as in K-theory, the three types coincide.
(2) Equivalence projections are also called Murray-von Neumann equivalence, and homotopy is sometimes called strong equivalence.
(3) Equivalence in commutative C^*-algebras is nothing but equality. If A is $B(H)$, then it is easy to prove that

$$p \sim q \Leftrightarrow dimpH = dimqH \Leftrightarrow pH \text{ is isometric isomorphic to } qH.$$

We next give some basic properties:

Proposition 2.12 *If $p_1,\ p_2$ and $q_1,\ q_2$ are projections in C^*-algebra A, and if $p_1 \sim q_1,\ p_2 \sim q_2$; and $p_1 \perp p_2;\ q_1 \perp q_2$, then $p_1 \oplus p_2 \sim q_1 \oplus q_2$.*

Proof Let u and v be in A such that $p_1 = u^*u,\ q_1 = uu^*$ and $p_2 = v^*v,\ q_2 = vv^*$. Take $w = u \oplus v$, then we have

$$p_1 \oplus p_2 = w^*w,$$

and

$$q_1 \oplus q_2 = ww^*.$$

Proposition 2.13 *The notations are the same as in Definition 2.9. If there exists an invertible element z in A such that $q = zpz^{-1}$, then $p \sim_u q$.*

Proof Since $zp = qz$ and $z^*q = pz^*$ by the assumption, it follows that

$$pz^*z = z^*qz = z^*zp,$$

that is, p commutes with z^*z. Because $|z|^{-1} = (z^*z)^{-1/2}$ can be uniformly approximated by a polynomial sequence with zero constant term about z^*z, so that p commutes with $|z|^{-1}$. Define a unitary operator $u = z|z|^{-1}$, then we have

$$\begin{aligned} upu^* &= z|z|^{-1}p|z|^{-1}z^* \\ &= zp|z|^{-2}z^* \\ &= qz|z|^{-2}z^* = q. \end{aligned}$$

Proposition 2.14 *Suppose that* p *and* q *are projections in* C^**-algebra* A*. Then* $p \sim_u q \Leftrightarrow p \sim q$ *and* $1 - p \sim 1 - q$.

Proof The necessity is obvious. Let us prove the sufficiency. From the assumption, there exist partial isometric operators v and w in A such that $v^*v = p,\ vv^* = q$ and $w^*w = 1 - p,\ ww^* = 1 - q$. It is easy to check that $vw^* = 0$, that is, $v \perp w$. Let $u = v \oplus w$. Hence u is unitary since $u^*u = v^*v + w^*w = 1$ and $uu^* = vv^* + ww^* = 1$. Also

$$vpw^* = wpv^* = wpw^* = 0.$$

Therefore

$$\begin{aligned} upu^* &= (v \oplus w)p(v^* \oplus w^*) \\ &= vpv^* = q. \end{aligned}$$

Proposition 2.15 *Suppose that* p *and* q *are projections in* C^**-algebra* A*. If* $||p - q|| < 1$ *then* $p \sim_h q$.

Proof Define symmetric elements $v_p = 2p-1$ and $v_q = 2q-1$. Let $z = v_q v_p + 1$, then $qz = zp$. From the assumption $||p - q|| < 1$, it implies that $||z - 2|| < 2$, which shows that z is invertible. Define a unitary operator $u = z|z|^{-1}$, then by direct calculation $q = upu^*$. Following the above process, let $z_t = tz + 2 - 2t$ for all $t \in [0, 1]$. Correspondingly, we obtain the unitary elements $u_t = z_t|z_t|^{-1}$ $t \in [0,\ 1]$. Since the mapping $x \to x|x|^{-1}$ $(x \in GL(A))$ is continuous, where $GL(A)$ is denoted as the set of all invertible elements in A, and the mapping $t \to z_t$ $(t \in [0,\ 1])$ is also. Therefore the mapping defined by $\phi : t \to u_t p u_t^*$ $(t \in [0,\ 1])$ is continuous, and we have that $\phi(0) = p$ and $\phi(1) = q$.

Corollary 2.3 *If p and q are projections in C^*-algebra A. Then*

$$p \sim_h q \Rightarrow p \sim_u q \Rightarrow p \sim q.$$

The inverse proposition of Corollary 2.3 is generally not true. But it can be proved that

$$p \sim q \Rightarrow \begin{pmatrix} p & 0 \\ 0 & 0 \end{pmatrix} \sim_u \begin{pmatrix} q & 0 \\ 0 & 0 \end{pmatrix},$$

and

$$p \sim_u q \Rightarrow \begin{pmatrix} p & 0 \\ 0 & 0 \end{pmatrix} \sim_h \begin{pmatrix} q & 0 \\ 0 & 0 \end{pmatrix}.$$

For details, please refer to Proposition 5.2.12 in [23].

Suppose that A is a C^*-algebra. Set $M_\infty = \cup_{n=1}^{\infty} M_n(A)$, the inductive limit of the direct sequence M_∞ is $A \otimes K(H)$. Write $\widetilde{P(A)}$ as the set of all projections in M_∞.

Definition 2.10 Let p and q be in $\widetilde{P(A)}$. Then p and q are said to be equivalent, write $p \sim q$, if there exists a v in $M_\infty(A)$ such that $p = v^*v$ and $q = vv^*$. Write its equivalence class of p as $\widetilde{p}$. The set of all equivalence class is denoted by $V(A)$.

projections

Remark 2.9 The three relations about projections are equivalent to each other on $\widetilde{P(A)}$.

The equivalence relation in $\widetilde{P(A)}$ has the following basic properties:

Proposition 2.16 *Let p, q and p′, q′ be in* $\widetilde{P(A)}$.

(1) If $p \sim p'$ and $q \sim q'$, then $p \oplus q \sim p' \oplus q'$;
(2) $p \oplus q \sim q \oplus p$;
(3) If $p,\ q \in M_n(A)$ and $pq = 0$, then $p + q \sim p \oplus q$.

Proof

(1) If $p = u^*u,\ p' = uu^*$ and $q = v^*v,\ q' = vv^*$. Set $w = u \oplus v$, then

$$p \oplus q = w^*w;\ p' \oplus q' = ww^*.$$

Thus $p \oplus q \sim p' + q'$.

(2) Let $u = \begin{pmatrix} 0 & q \\ p & 0 \end{pmatrix}$. Hence $p \oplus q = u^*u$ and $q \oplus p = uu^*$.

(3) Note that if $p \in M_n(A)$, then $p \sim p \oplus 0_m$, where 0_m is the $m \times m$ matrix all of whose entries are zero. Indeed, let $u = (p, 0_{nm})$, then $u^*u = p \oplus 0_m$ and

$uu^* = p$. Using the supposition $p\,q = 0$, set $v = \begin{pmatrix} p & q \\ 0_n & 0_n \end{pmatrix}$, it is easy to check that $v^*v = p\oplus q$ and $vv^* = (p+q)\oplus 0_n$. Hence, $p+q \sim (p+q)\oplus 0_n \sim p\oplus q$.

Remark 2.10 If $p,\ q \in \widetilde{P(A)}$, identifying p with $\begin{pmatrix} p & 0 \\ 0 & 0 \end{pmatrix}$ (denoted by $diag(p,\ 0)$) and q with $\begin{pmatrix} q & 0 \\ 0 & 0 \end{pmatrix}$ (denoted by $diag(q,\ 0)$), then

$$p = diag(p,\ 0) \sim diag(0,\ p) \perp diag(q,\ 0) = q.$$

Now we introduce the addition operation on $V(A)$ as follows.

$$\widetilde{p} + \tilde{q} = \widetilde{diag(p+q)},$$

for all $\widetilde{p}$ and $\widetilde{q}$ in $V(A)$. From Remark 2.9, we see that for all p and q in $\widetilde{P(A)}$ there exist p' and q' in $\widetilde{P(A)}$, such that $p' \sim p$ and $q' \sim q$ and satisfying $p' \perp q'$. It follows that

$$\widetilde{p} + \widetilde{q} = \widetilde{diag(p+q)} = \widetilde{p' \oplus q'},$$

combining with Proposition 2.16. The zero element is denoted as $\widetilde{0_n}$, in which n is any natural number. Because $\widetilde{0_m} \sim \widetilde{0_n}$, therefore the zero element can be simply denoted as $\widetilde{0}$. According to Proposition 2.16, the addition defined in $V(A)$ satisfies the associative law and the commutative law. So $V(A)$ becomes a commutative semigroup with respect to this addition.

Definition 2.11 Let A be a C^*-algebra. Then the Grothendieck universal group of $V(A)$ is called K_0 group of A, which is denoted by $K_0(A)$.

For characterizing K_0-group, we briefly introduce the Grothendieck universal group of abelian semigroup. Let M be a commutative semigroup with zero element. Then commutative group $G(M)$ is called the Grothendieck universal group of M, or universal group of M in short, if there exists a homomorphism mapping $\phi : M \to G(M)$ with the universal property, that is, if for any commutative group W and homomorphism mapping ψ from M to W, then there exists uniquely homomorphism $\widetilde{\psi} : G(M) \to W$ such that $\widetilde{\psi} \circ \phi = \psi$, namely, the following exchange chart holds (Fig. 2.1):

It is easy to see that if the universal group of commutative semigroups exists, then it is unique in the sense of isomorphism. The following proposition shows that the universal group of commutative semigroups must exist, and the construction method is given.

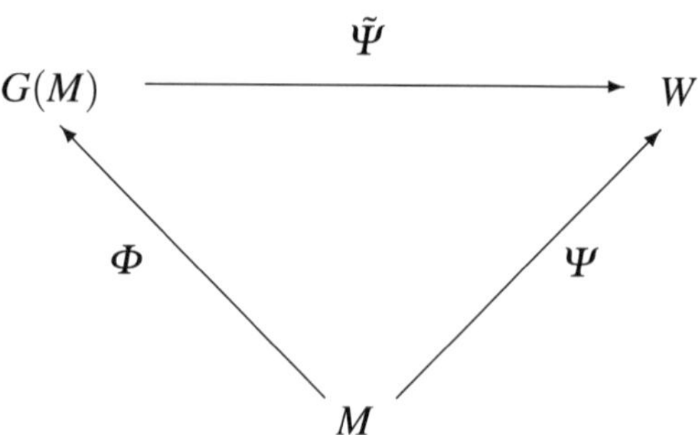

Fig. 2.1 Exchange chart

Proposition 2.17 *The universal group of commutative semigroups M is unique in the sense of isomorphism.*

Proof We need only to prove the existence, because the uniqueness has been proved above. Define an equivalence relation $\sim$ on $M \times M$ by setting $(x, \ y) \sim (z, \ t)$, if there exists an element u in M such that $x+t+u = z+y+u$. Denote by $[x, \ y]$ the equivalence class of $(x, \ y)$. Let $G(M) = (M \times M)/\sim$ be the set of all equivalence classes. Define addition operation on $G(M)$ given by

$$[x, \ y] + [z, \ t] = [x + z, \ y + t].$$

It shows that $G(M)$ has zero element $[0, \ 0] = [x, \ x]$, and every element $[x, \ y]$ has an inverse element $[y, \ x]$. Thus $G(M)$ is a commutative group. Define a mapping ϕ from M to $G(M)$ given by $x \to [x, 0]$. It is easy to check that ϕ is a homomorphic mapping. Since

$$[x, \ y] = [x, \ 0] + [0, \ y]$$

$$= [x, \ 0] - [y, \ 0],$$

$G(M) = \phi(M) - \phi(M)$. Next, we want to prove that $G(M)$ is the universal group of M. For any commutative group G and homomorphism $\psi : M \to G$, let

$$\widetilde{\psi} : G(M) \to G, \widetilde{\psi}([x, \ y]) = \psi(x) - \psi(y) \ ([x, \ y] \in G(M)).$$

From the construction of $\widetilde{\psi}$, we have $\widetilde{\psi} \circ \phi = \psi$.

Now it is necessary to show the rationality and uniqueness of the definition of $\widetilde{\psi}$ as follows. In fact, if $(x_1, \ x_2) \sim (y_1, \ y_2)$, then there exists a $u \in M$ such that $x_1 + y_2 + u = y_1 + x_2 + u$. Therefore,

$$\psi(x_1) + \psi(y_2) + \psi(u) = \psi(y_1) + \psi(x_2) + \psi(u),$$

that is, $\psi(x_1)-\psi(x_2)=\psi(y_1)-\psi(y_2)$. It implies that $\widetilde{\psi}([x_1,\ x_2])=\widetilde{\psi}([y_1,\ y_2])$, which shows that $\widetilde{\psi}$ is well defined. Finally, if there is another homomorphism $\widetilde{\psi}'$: $G(M)\to G$ such that $\widetilde{\psi}'\circ\phi=\psi$, then we have

$$\begin{aligned}\widetilde{\psi}([x,\ y]) &= \psi(x)-\psi(y)\\ &= \widetilde{\psi}'\phi(x)-\widetilde{\psi}'\phi(y)) = \widetilde{\psi}'([x,\ y]).\end{aligned}$$

Thus $\widetilde{\psi}=\widetilde{\psi}'$.

Remark 2.11

(1) $\phi : M \to G(M)$ is injective if and only if M satisfies the cancellative law. (referring to [14]);
(2) The above semigroup $V(A)$ satisfies cancellative law by using Theorem 7.1.2 in [17].

Proposition 2.18 *We write* $[p]$ *for the image of* $\widetilde{p}$ *in* $K_0(A)$. *Then*

(1) $K_0(A)=\{[p]-[q] : p, q\in\widetilde{P(A)}\}=\{[p]-[1_n] :\ p\in\widetilde{P(A)},\ n\in\mathbb{N}\}$;
(2) If $p,\ q\in\widetilde{P(A)}$, *then* $[p]=[q]$ *in* $K_0(A)$ *if and only if there exists an element* r *in* $\widetilde{P(A)}$ *such that* $p\oplus r\sim q\oplus r$ *in* $\widetilde{P(A)}$. *Indeed,* r *can be selected as the form* 1_n, *where* 1_n *is the unit of* $M_n(A)$, *namely, the* $n\times n$ *matrix all of whose entries are 0, except for those on the main diagonal.*

Proof

(1) We first have $K_0(A)=\{[p]-[q] :\ p,\ q\in P(A)\}$ from Proposition 2.17. If $q\in M_n(A)$ then

$$\begin{aligned}[p]-[q] &= [p]+[1_n-q]-([q]-[1_n-q])\\ &= [p\oplus(1_n-q)]-[1_n].\end{aligned}$$

Hence $K_0(A)=\{[p]-[1_n] : p\in\widetilde{P(A)},\ n\in\mathbb{N}\}$.
(2) From the construction of universal group, we see that $[p]-[0]=[q]-[0]$ if and only if $(\widetilde{p},\ \widetilde{0})\sim(\widetilde{q},\widetilde{0})$ in $V(A)\times V(A)$. That is, there exists $r\in P(A)$ such that $\tilde{p}+\tilde{r}=\tilde{q}+\tilde{r}$ in $V(A)$. Equivalently, $p\oplus r\sim q\oplus r$ in $\widetilde{P(A)}$.

Remark 2.12

(1) Projections p and q in $\widetilde{P(A)}$ are called stable equivalence if there exists $1_n\in M_n(A)$ such that $1_n\oplus p\sim 1_n\oplus q$ in $\widetilde{P(A)}$. We denote by $p\approx q$ in case p and q are stable equivalence, and by $[p]$ the stable equivalence class of p. According to Proposition 2.18, $[p]$ the stable equivalence class of p and the image of $\widetilde{p}$ in $K_0(A)$ are the same. That is why we use the same symbol to represent two different things in form, because they are actually the same.

(2) Suppose that A and B are C^*-algebras. If $\phi : A \to B$ is a $*$ homomorphism, then $\phi^n : M_n(A) \to M_n(B)$ is also. If p and q are projections in $M_n(A)$ and $p \sim q$, then $\phi^n(p) \sim \phi^n(q)$. Moreover, if $p \oplus 1_{M_n(A)} \sim q \oplus 1_{M_n(A)}$, then

$$\phi^n(p) \oplus \phi^n(1_{M_n(A)}) \sim \phi^n(q) \oplus \phi^n(1_{M_n(A)}),$$

where $1_{M_n(A)}$ is the above defined unit 1_n. Here, we emphasize that $1_{M_n(A)}$ is the unit of $M_n(A)$. It follows that

$$\phi^n(p) \oplus \phi^n(1_{M_n(A)}) \oplus (1_{M_n(B)} - \phi^n(1_{M_n(A)}))$$

$$\sim \phi^n(q) \oplus \phi^n(1_{M_n(A)}) \oplus (1_{M_n(B)} - \phi^n(1_{M_n(A)})),$$

that is, $\phi^n(p) \oplus 1_{M_n(B)} \sim \phi^n(q) \oplus 1_{M_n(B)}$. The above proof shows that if $p \approx q$, then $\phi^n(p) \approx \phi^n(q)$. Hence, ϕ can induce $*$ homomorphism $\phi_* : \widetilde{P(A)} \to \widetilde{P(B)}$ such that $\phi_*(\widetilde{p}) = \widetilde{\phi(p)}$. Therefore, there exists a unique homomorphism; we write still as ϕ_*, from $K_0(A)$ to $K_0(B)$ such that $\phi_*([p]) = [\phi(p)]$ by the universality of K_0-group.

Furthermore, for given C^*-algebras A, B and C, if there exist unital $*$ homomorphisms $\varphi : A \to B$ and $\psi : B \to C$, then $(\psi\varphi)_* = \psi_*\varphi_*$ and $(id_A)_* = id_{K_0(A)}$. Hence, we have a covariant functor

$$A \to K_0(A), \ \varphi \to \varphi_*,$$

from the category of all unital C^*-algebras to the category of K_0 groups.

Example 2.3 Let A be the complex field $\mathbb{C}$ and projections p and q in $\widetilde{P(\mathbb{C})}$. Then there exists a natural number n such that $p, \ q \in M_n(\mathbb{C})$. It is easy to check that

$$p \sim q \sim\Leftrightarrow rank(p) = rank(q).$$

Since $rank(p \oplus q) = rank(p) + rank(q)$, we can define a homomorphism $\phi : \widetilde{P(\mathbb{C})} \to \mathbb{Z}$ such that $\phi(\tilde{p}) = rank(p)$. Therefore there exists uniquely homomorphism; we write still as ϕ, from $K_0(A)$ to $\mathbb{Z}$ such that $\phi([p]) = rank(p)$,; it is obviously surjective since $\phi([1]) = 1$, where number 1 is the generated element of integers set $\mathbb{Z}$.

We next want to show that ϕ is injective. Assume that $x \in ker\phi$,, there exists $p, \ q \in M_n(\mathbb{C})$ such that $x = [p] - [q]$ and

$$\phi(x) = \phi([p]) - \phi([q]) = rank(p) - rank(q) = 0,$$

which shows that p and q have the same rank in $M_n(\mathbb{C})$. Thus $p \sim q$ in $M_n(\mathbb{C})$, so $[p] = [q]$. Hence $x = 0$ from the above proof, which shows that ϕ is injective. Therefore ϕ is an isomorphism.

From the above observation, we see that K_0 group is regarded as the "dimension" group, so that $[p]$ is the generalized dimension of p.

It is parallel to the algebra K-theory; the above $K_0(A)$ group of C^*-algebra can also be characterized by finitely generated projective modules. This is just the starting point of the generalized Fredholm index theory.

Definition 2.12 Let A be a unital C^*-algebra, and let $P[A]$ be the set of all algebraically finitely generated projections over A. Taking unitary equivalence $\cong$ as the equivalence relation, and the direct sum $\oplus$ as addition operation, then the set $(P[A]/\cong, \oplus)$ of all equivalence classes is a commutative semigroup, which is denoted as $V[A]$.

Note that for a finitely generated Hilbert C^*-module, the unitary equivalence as a Hilbert C^*-module is identical with the module isomorphism as a Banach- A-module (refer to Theorem 3.1 in [10]). Parallel to Proposition 4.1.17 in [14], we can obtain the following conclusion.

Proposition 2.19 *The notations are as the same in the above. Then we have* $V(A) \cong V[A]$.

From Proposition 2.19 combined with Proposition 2.18, we have

$$K_0(A) = \{[P] - [Q] : \ P,\ Q \in P[A]\} = \{[P] - [A^n] : \ P \in P[A],\ n \in \mathbb{N}\}.$$

2.3.3 *Fredholm Modules Mappings and Their Generalized Index*

Definition 2.13 Let E be a Hilbert A-module and $\pi : L(E) \to L(E)/K(E)$ be the quotient homomorphism. Then an element T in $L(E)$ is called a Fredholm module operator if $\pi(T)$ is invertible in $L(E)/K(E)$.

The set of all Fredholm module operators in $L(E)$ is denoted by $\Phi(E)$. Therefore $\Phi(E) = \{T \in L(E) : \pi(T) \in Inv(L(E)/K(E))\}$ from Definition 2.13, where $Inv(L(E)/K(E))$ is the set of all invertible elements in $L(E)/K(E)$.

As an extreme case, if A is a unital C^*-algebra, then as Hilbert A-module itself we have $L(E) = K(A) = A$. Thus $L(A)/A = [0]$. This shows that there is no Fredholm module operator in $L(A)$ by Definition 2.13. Therefore, it is of no practical significance to discuss Fredholm module operators generally.

In what follows, the theory of Fredholm module operators is developed on $H \otimes A$, where A is a unital C^*-algebra and H is a separable Hilbert space. We also have to look for ideas from the Fredholm operators on Hilbert space. We see that Hilbert space can be regarded as a Hilbert module over the complex field $\mathbb{C}$. The index set of $\Phi(H)$ is the integer set $\mathbb{Z}$, which is exactly the K_0-group of complex field $\mathbb{C}$. Therefore, we hope that the index of Fredholm module operators belongs to $K_0(A)$ is a natural imagination.

It is known from Sect. 2.3.1 that for a classical Fredholm operator T on Hilbert space, its range $R(T)$ is of course a closed subspace, and $ker\ T$ and $ker\ T^*$ are finite dimensional subspaces. But for the Fredholm module operator T defined above, its range $R(T)$ is usually not a closed submodule, and $ker\ T$ and $ker\ T^*$ as A-modules are generally not finitely generated. Therefore, in order to extend the classical Fredholm index theory to Hilbert module framework, we need a lot of preparation and overcome great difficulties as follows.

In order to introduce the index on elements in $\Phi(H \otimes A)$, we give the Atkinson type theorem in the following.

Theorem 2.4 (Atkinson type theorem) *Let T be in $L(H \otimes A)$. Then the following conclusions are equivalent:*

(1) $T \in \Phi(H \otimes A)$;
(2) There exists an element G in $L(H \otimes A)$ with $R(G)$ closed, and $ker\ G$ and $ker\ G^*$ are algebraically finitely generated projective modules in $H \otimes A$, such that $T - G \in K(H \otimes A)$.

Remark 2.13 G is called a compact perturbation of T.

We need the following lemma.

Lemma 2.4 *There exists a bijection between the set of all algebraically finitely generated projective submodules in $H \otimes A$ and the set of all projections in $K(H \otimes A)$. Concretely, for any finitely generated projective submodule F in $H \otimes A$, there exists a unique projection P_F in $K(H \otimes A)$ such that $R(P_F) = F$. On the contrary, for any projection P in $K(H \otimes A)$, then its range $R(P)$ is an algebraically finitely generated projective submodule in $H \otimes A$.*

Proof Choose the canonical orthonormal basis of H given by $e_1 = (1,\ 0,\ \cdots,\ 0,\ \cdots)$, $e_2 = (0,\ 1,\ 0,\ \cdots,\ 0,\ \cdots)$, $\cdots$, $e_n = (0,\ 0,\ \cdots,\ 0,\ 1,\ 0,\ \cdots)$, $\cdots$. So we get the orthogonal basis $\{e_n \otimes 1\}_{n=1}^{\infty}$ of $H \otimes A$, which is also denoted as $\{e_n\}$ without causing confusion later. Let $P_n = \sum_{k=1}^{n} \theta_{e_k,\ e_k} \in K(H \otimes A)$. Hence every P_n is projection, and it is easy to check that $\{P_n\}_{n=1}^{\infty}$ is an approximate unit in $K(H \otimes A)$. Thus for any projection $P \in K(H \otimes A)$, there exists natural number n such that $||(1 - P_n)P|| < \frac{1}{12}$. Set $Q_n = P_n P P_n$, then Q_n is a positive element in $K(H \otimes A)$ and $||Q_n|| \leq 1$. Therefore

$$
\begin{aligned}
||Q_n^2 - Q_n|| &\leq ||Q_n(Q_n - P)|| + ||(Q_n - P)P|| + ||P - Q_n|| \\
&< 3 \times \frac{1}{12} = \frac{1}{4},
\end{aligned}
$$

which shows that $\sigma(Q_n) \subset [0,\ \frac{1}{2}) \cup (\frac{1}{2},\ 1]$. Define a function given by

$$
f(t) = \begin{cases} 0, \text{ if } t < \frac{1}{2}, \\ 1, \text{ if } t \geq \frac{1}{2}. \end{cases}
$$

From the construction of f, we see that it is continuous when restricted to $\sigma(Q_n)$. Thus $f(Q_n)$ is a projection and $||Q_n - f(Q_n)|| < \frac{1}{2}$ by the function calculus of positive elements in $K(H \otimes A)$. It follows that

$$\begin{aligned} ||P - f(Q_n)|| &\leq ||P - Q_n|| + ||Q_n - f(Q_n)|| \\ &< \frac{1}{12} + \frac{1}{2} < 1. \end{aligned}$$

Hence there exists a unitary u in $L(H \otimes A)$ such that $uPu^* = f(Q_n)$ by applying Proposition 2.15. Therefore

$$R(P) = P(H \otimes A) \cong R(f(Q_n)) = f(Q_n)(H \otimes A),$$

is a closed submodule in $H \otimes A$. Since $f(t)$ can be uniformly approximated on $\sigma(Q_n)$ by a polynomial sequence with constant terms being zero, $P_n f(Q_n) = f(Q_n)P_n = f(Q_n)$, from which it implies that $f(Q_n) \leq P_n$. Note that $P_n(H \otimes A) \cong A^n$, thus

$$f(Q_n)(H \otimes A) \oplus (P_n - f(Q_n))(H \otimes A) \cong A^n.$$

Therefore $R(f(Q_n)) = f(Q_n)(H \otimes A)$ is an algebraically finitely generated projective submodule in $H \otimes A$.

Conversely, if F is an algebraically finitely generated projective submodule in $H \otimes A$, then there exists Hilbert A-module E and a natural number n such that $F \oplus E \cong A^n$, which follows that $E \oplus (H \otimes A) \cong H \otimes A$ by the Kasparov's stabilization theorem. Therefore there exists projection $P_E \in L(H \otimes A)$ such that $R(P_E) \cong E$. Now we define a projection operator $P_F \in L(H \otimes A)$ that is from $H \otimes A$ onto F and then consider the projection $P_F \oplus P_E$ on $(H \otimes A) \oplus (H \otimes A)$. From the above proof, we see that $R(P_F \oplus P_E) \cong A^n$. Next, take $\{\xi\}_{i=1}^n$ to be a standard orthogonal basis of $R(P_F \oplus P_E)$. Thus $P_F \oplus P_E = \sum_{i=1}^n \theta_{\xi_i, \xi_i}$. Therefore

$$P_F \oplus P_E \in K((H \otimes A) \oplus (H \otimes A)).$$

So that P_F is in $K(H \otimes A)$.

Remark 2.14 From the proof of Lemma 2.4, it can be seen that algebraically finitely generated closed submodule must be projective module.

Corollary 2.4 *Let T be in $L(H \otimes A)$ with closed range. If $kerT$ and $kerT^*$ are algebraically finitely generated projective modules, then T is a Fredholm module operator.*

Proof Since $R(T)$ is closed, $T = V|T|$ by Proposition 1.9, where $V \in L(H \otimes A)$ is a partial isometry satisfying $R(V) = R(T)$ and $kerV = kerT$. Let P be the projection from $H \otimes A$ onto $kerT$; it follows from Lemma 2.4 that P is in $K(H \otimes A)$.

It is easily checked that $|T| + P$ is invertible. Therefore $\pi(|T|) = \pi(|T| + P)$ is invertible in $L(H \otimes A)/K(H \otimes A)$. Since $R(I - V^*V) = ker V$ and $R(I - VV^*) = ker V^*$ are algebraically finitely generated closed submodules, by using Lemma 2.4 again, it shows that $I - V^*V$ and $I - VV^*$ belong to $K(H \otimes A)$. Thus $\pi(V)$ is unitary in $L(H \otimes A)/K(H \otimes A)$, from which it implies that $\pi(T) = \pi(V)\pi(|T|)$ is invertible in $L(H \otimes A)/K(H \otimes A)$. So that T is a Fredholm module operator in $L(H \otimes A)$ by Definition 2.13.

The Proof of Theorem 2.4. (2) $\Rightarrow$ (1) Suppose that T is in $L(H \otimes A)$, and there exists G in $L(H \otimes A)$ with closed range $R(G)$ and satisfying that $ker G$ and $ker G^*$ are algebraically finitely generated closed submodules, such that $T - G \in K(H \otimes A)$. Then by Lemma 2.4 that G is in $\Phi(H \otimes A)$. Hence $\pi(T) = \pi(G)$ is invertible in $L(H \otimes A)/K(H \otimes A)$; this shows that $T \in \Phi(H \otimes A)$ by Definition 2.13.

(1) $\Rightarrow$ (2) If $T \in \Phi(H \otimes A)$, then $\pi(T)$ is invertible by Definition 2.13; hence $\pi(|T|)$ is also invertible. Since π is surjective, there exists $U \in L(H \otimes A)$ such that $\pi(U) = \pi(T)\pi(|T|)^{-1}$. It is easy to check that $\pi(U)$ is unitary in $L(H \otimes A)/K(H \otimes A)$. Next, we shall prove that there exists a partial isometry $V \in L(H \otimes A)$ such that $\pi(V) = \pi(U)$. If this is proved, then V is the compact perturbation. Let $(P_n)_{n=1}^{\infty}$ be the approximation unit element of $K(H \otimes A)$ constructed in Lemma 2.4. Note that $I - U^*U \in K(H \otimes A)$,, then there exists a sufficiently large natural number n such that $||(I - P_n)(I - U^*U)(I - P_n)|| < 1$. Hence

$$(I - P_n) - (I - P_n)(I - U^*U)(I - P_n)$$

$$= (I - P_n)U^*U(I - P_n),$$

is invertible in $(I - P_n)L(H \otimes A)(I - P_n)$. Because P_n is the unit element in $P_n + (I - P_n)U^*U(I - P_n)$, it is reversible in $P_n + (I - P_n)U^*U(I - P_n)$. Therefore $P_n + (I - P_n)U^*U(I - P_n)$ is invertible in $L(H \otimes A)$. Let $V = U(I - P_n)[P_n + (I - P_n)U^*U(I - P_n)]^{-1/2}$, then by routine calculation, $V^*V = I - P_n$ and $\pi(V) = \pi(U)$. Thus V is a partial isometry. Choose a self-adjoint element $H \in L(H \otimes A)$ such that $\pi(H) = \ln \pi(|T|)$. Set $G = Ve^H$. Since

$$\pi(G) = \pi(V)\pi(e^H) = \pi(V)e^{\pi(H)}$$

$$= \pi(U)\pi(|T|) = \pi(T),$$

G is a compact perturbation of T. Because the positive element e^H is invertible in $L(H \otimes A)$, $R(G) = R(V)$ is a closed submodule, and $ker G$ is algebraically finitely generated since $ker G = ker V = R(P_n)$ by the above proof. Furthermore, since

$$\pi(I - VV^*) = \pi(I) - \pi(V)\pi(V^*)$$

$$= \pi(I) - \pi(U)\pi(U^*) = 0,$$

$I - VV^* \in K(H \otimes A)$. Hence by Lemma 2.4 that $R(I - VV^*)$ is an algebraically finitely generated closed, so is $kerG^*$ since $kerG^* = kerV^* = R(I - VV^*)$.

Theorem 2.4 indicates that each Fredholm module operator has compact perturbation. But in general, the compact perturbation is not unique. This brings new problems to the definition of generalized Fredholm index, that is, for a given $T \in \Phi(H \otimes A)$, if G_1 and G_2 are compact perturbations of T, then whether or not is $[kerG_1] - [kerG_1^*] = [kerG_2] - [kerG_2^*]$?

The answer is yes; this is one of the key technologies that G.G.Kasparov [13] solved when he generalized the theory of classical Fredholm operators. For details, please refer to Section 17.2 of [23].

Definition 2.14 Let T be in $\Phi(H \otimes A)$, and let G be a compact perturbation of T. Define the generalized Fredholm index of T given by

$$Index(T) = [kerG] - [kerG^*] \in K_0(A).$$

Remark 2.15

(1) It is easily checked by Definition 2.14 that

$$Index(T \oplus S) = Index(T) + Index(S),$$

for all $T,\ S \in \Phi(H \otimes A)$.

(2) Suppose that $T \in \Phi(H \otimes A)$. If G is a compact perturbation of T, then there exists polar decomposition $G = V|G|,\ V \in L(H \otimes A)$ is a partial isometry, and $kerV = kerG,\ kerV^* = kerG^*$. It follows that $R(I - V^*V) = kerG$ and $R(I - VV^*) = kerG^*$. Since

$$I - V^*V \sim I - VV^* \Leftrightarrow R(I - V^*V) \cong R(I_V V^*),$$

and therefore $Index(T) = [I - V^*V] - [I - VV^*]$.

In the following, we shall characterize the properties of the generalized Fredholm index.

Proposition 2.20 *Suppose that $F \in \Phi(H \otimes A)$. If $K \in K(H \otimes A)$, then $F + K \in \Phi(H \otimes A)$ and $Index(F + K) = Index(F)$. Moreover, if $T \in L(H \otimes A)$ is invertible, then $TF,\ FT \in \Phi(H \otimes A)$ and $Index(TF) = Index(FT) = Index(F)$.*

Proof Since $\pi(F + K) = \pi(F)$ is invertible in $L(H \otimes A)/K(H \otimes A)$, $F + K \in \Phi(H \otimes A)$. Hence by Theorem 2.4, there exists a compact perturbation G of F such that $F - G \in K(H \otimes A)$. It follows that G is also a compact perturbation of $F + K$ since $F + K - G = (F - G) + K \in K(H \otimes A)$. Therefore by Definition 2.14,

$$Index(F) = Index(F + K) = Index(G).$$

Finally, if T is invertible, then TG and GT are compact perturbations of TF and FT, respectively. We consider only the case of TF, and the case of FT can be obtained similarly. Since $kerTG = kerG$ and $ker(TG)^* = kerG^*T^* \approx kerG^*$. It follows by Definition 2.14 that

$$Index(F) = [kerG] - [kerG^*]$$

$$[ker(TG)] - [ker(TG)^*]$$

$$= Index(TF).$$

Proposition 2.21 *If* $F \in \Phi(H \otimes A)$ *and* $Index(F) = 0$, *then there exists an invertible element* $G \in L(H \otimes A)$ *such that* $F - G \in K(H \otimes A)$.

Proof Without loss of generality, suppose that F has closed range $R(F)$, and $kerF$ and $kerF^*$ are algebraically finitely generated closed submodules. Let $F = V|F|$ be its polar decomposition of F. Since $Index(F) = 0$, $P_n(H \otimes A)$ and $kerF^*$ are stably equivalent, in which identifying $kerF$ with $P_n(H \otimes A)$. Hence there exists an algebraically finitely generated projective module E such that

$$P_n(H \otimes A) \oplus E \cong kerF^* \oplus E.$$

Thus by Lemma 2.4, there exists a projection $P \in K(H \otimes A)$ such that $P(H \otimes A) \cong E$ and $P \leq P_{n+m} - P_n$. Set $V_1 = V(I - P)$, then V_1 is a partial isometry, and we have

$$kerV_1 = P_n(H \otimes A) \oplus P(H \otimes A),$$

and

$$kerV^* = kerV^* \oplus V(E) \cong kerF^* \oplus E.$$

Therefore there exists a partial isometry $W \in K(H \otimes A)$ such that $W^*W = kerV_1$ and $WW^* = kerV_1^*$. Let $G = (V_1 + W)(|F| + P_n),$; it follows from what we have just shown that G is invertible and $F - G \in K(H \otimes A)$.

With the above proposition, next we shall prove the continuity of the index mapping.

Theorem 2.5 *The index mapping*

$$Index : \Phi(H \otimes A) \to K_0(A),\ T \to Index(T)$$

is a continuous homomorphism.

Proof First of all, $\Phi(H \otimes A)$ is equipped with quotient topology, and $K_0(A)$ is endowed with discrete topology. Therefore, to prove that $Index$ is continuous, we need only prove that $Index$ is a local constant, that is to say that $Index$ is invariant under small perturbation. For any $F \in \Phi(H \otimes A)$, since the mapping from $L(H \otimes A)$ to $L(H \otimes A)/K(H \otimes A)$ given by

$$T \to \pi(T)\pi(|T|)^{-1} \ (T \in L(H \otimes A))$$

is continuous, it follows that there exists sufficiently small $\varepsilon > 0$ such that if $G \in \Phi(H \otimes A)$ and $||F - G|| < \varepsilon$, then

$$||\pi(F)\pi(|F|)^{-1} - \pi(G)\pi(|G|)^{-1}|| < 1.$$

In order to prove the continuity of index mapping, it is necessary to prove $Index(F) = Index(G)$. Choose U and V in $L(H \otimes A)$ satisfying $||U|| < \sqrt{2}$ and $||U - V|| < \sqrt{2}$ such that $\pi(U) = \pi(F)\pi(|F|)^{-1}$ and $\pi(V) = \pi(G)\pi(|G|)^{-1}$. Next, we want to prove that $Index(F) = Index(U)$. $Index(G) = Index(V)$ can be similarly obtained. Let F_1 be a compact perturbation of F; hence $\pi(F_1) = \pi(F)$,; it follows that $\pi(|F_1|) = \pi(|F|)$. Thus $\pi(U) = \pi(F_1)\pi(|F_1|)^{-1}$. Let $F_1 = V_1|F_1|$ be the polar decomposition of F_1, where V_1 is a partial isometry. It implies that $\pi(U) = \pi(V_1)$, so that $Index(U) = Index(V_1)$. Note that $Index(V_1) = Index(F_1)$, hence $Index(U) = Index(F_1) = Index(F)$. Further, let $X = 2I - U^*U + U^*V$ and $Y = 2I - UU^* + UV^*$. Then it is easy to check that $||2I - X|| < 2$ and $||2I - Y|| < 2$, where I is the identity operator on $H \otimes A$, which shows that X and Y are invertible. Therefore by Proposition 2.20, we have $Index(UX) = Index U$ and $Index(YV) = Index V$. Since

$$UX - YV = (I - UU^*)(U - V) + U(I - VV^*) - (I - U^*U)V \in K(H \otimes A).$$

By Proposition 2.20 again, we have $Index(UX) = Index(YV)$,; it follows from the above proof that $Index(F) = Index(G)$.

Finally, to show that $Index$ a homomorphic mapping, for this, we need to prove that for any $F, G \in \Phi(H \otimes A)$ the following equation holds;

$$Index(FG) = Index(F) + Index(G).$$

Since

$$Index(F \oplus G) = Index(F) + Index(G),$$

and

$$Index(FG \oplus I) = Index(FG).$$

Therefore from the proof of the first part, as long as it is to prove that there is a norm continuous path consisted of Fredholm module operators connecting $F \oplus G$ and $FG \oplus I$, because the path can be divided into segments with sufficiently small distance between adjacent endpoints, so that the Fredholm module operators corresponding to the adjacent endpoints have the same index. Indeed, let

$$\phi(t) = \begin{pmatrix} F & 0 \\ 0 & I \end{pmatrix} \begin{pmatrix} \cos t & -\sin t \\ \sin t & \cos t \end{pmatrix} \begin{pmatrix} I & 0 \\ 0 & G \end{pmatrix} \begin{pmatrix} \cos t & \sin t \\ -\sin t & \cos t \end{pmatrix} \left(t \in \left[0, \frac{\pi}{2}\right]\right).$$

Then for every $t \in [0, \frac{\pi}{2}]$ that $\pi(\phi(t))$ is invertible in $L(H \otimes A \oplus H \otimes A)/K(H \otimes A \oplus H \otimes A)$. Hence $\phi(t) \in \Phi(H \otimes A \oplus H \otimes A)$ $(t \in [0, \frac{\pi}{2}])$. It follows from the above construction that

$$\phi(0) = \begin{pmatrix} F & 0 \\ 0 & G \end{pmatrix} = F \oplus G,$$

and

$$\phi\left(\frac{\pi}{2}\right) = \begin{pmatrix} FG & 0 \\ 0 & I \end{pmatrix} = FG \oplus I,$$

which shows that $\phi(t), t \in [0, \frac{\pi}{2}]$ is a continuous path connecting $F \oplus G$ and $FG \oplus I$. The proof is over.

Suppose that F is in $\Phi(H \otimes A)$. We write $[F]$ for the path component of F and set $[\Phi(H \otimes A)] = \{[F] : F \in \Phi(H \otimes A)\}$. Then $[\Phi(H \otimes A)]$ is a semigroup when endowed with the following multiplication

$$[F][G] = [FG] \; (F, G \in \Phi(H \otimes A)).$$

Furthermore, a stronger conclusion can be obtained.

Theorem 2.6 *$[\Phi(H \otimes A)]$ is a group with the above defined multiplication, and the index mapping*

$$Index : [\Phi(H \otimes A)] \to K_0(A), \; [T] \to Index(T)$$

is a surjective isomorphism.

In order to prove Theorem 2.6, we need to use the following Kuiper-Mingo-Cuntz-Higson theorem which is the key technical result of J.A. Mingo [15] refining Kasprov' generalized index theory.

Proposition 2.22 *If A is a σ-unital C^*-algebra, then the unitary group $M(K \otimes A) = L(H \otimes A)$ is contractible with respect to the norm topology. That is to say, for any unitary operator $U \in L(H \otimes A)$, there exists a norm continuous path consisting of unitary operators connecting U and I.*

J.A. Mingo [15] obtained the above result in the case that A is a unital C^*-algebra. The case where A is σ-$unital$ C^*-algebra is obtained by J. Cuntz and N. Higson in [7]. In particular, when A reduces to the complex field $\mathbb{C}$, it is the classical conclusion of Kuiper in 1965, that is, the unitary group of $B(H)$ acting on a separable space H is contractible when endowed with the norm topology. The proof of Proposition 2.22 can be referred to Chapter 16 in [23].

For proving Theorem 2.6, we need also the following lemma:

Lemma 2.5 *If projection P is in $K(H \otimes A)$, then $I - P \sim I$ in $L(H \otimes A)$, where I is the identity mapping on $H \otimes A$.*

Proof It suffices to prove that there is an isometrical module operator $V \in L(H_A)$ such that $P = I - VV^*$. Since $H \otimes A \cong l_2 \otimes A = l_2(A)$, $L(H \otimes A) \cong L(l_2(A))$ and $K(H \otimes A) \cong K(l_2) \otimes A$. Based on this observation, next we use $l_2(A)$ instead of $H \otimes A$. Let $\{e_{ij}\}$ be a set of matrix units for $K(l_2)$. Let $P_n = \sum_{i=1}^{n} e_{ii} \otimes 1 \in K(l_2) \otimes A, n = 1, 2, \cdots$. Then $\{P_n\}_{n=1}^{\infty}$ is an approximate unit of $K(l_2) \otimes A$, where 1 is the unit of A. Consider the case of $P \leq P_n$ first, and then deal with the general case at the end. In this case, we have $P \in M_n(A) \cong M_n(\mathbb{C}) \otimes A$. Hence there exists positive integer n and $a_{ij} \in A, i, j = 1, 2, \cdots, n$, such that $P = \sum_{i,j=1}^{n} e_{ij} \otimes a_{ij}$. Construct a unitary U from l_2 to $l_2 \otimes \mathbb{C}^n$ given by $Ue_{ij}U^* = e_{11} \otimes f_{ij}, i, j = 1, 2, \cdots, n$, where $\{f_{ij}\}$ is the set of matrix units for $K(l_2 \otimes \mathbb{C}^n)$ (Indeed, if $\{e_k\}_{k=1}^{\infty}$ is the canonical orthogonal basis of l_2, then the unitary operator U from l_2 to $l_2 \otimes \mathbb{C}^n$ mapping $span\{e_1, e_2, \cdots, e_n\}$ onto $span\{e_1\} \otimes \mathbb{C}^n$ meets the above requirements). Let $p' = \sum_{i,j=1}^{n} f_{ij} \otimes a_{ij} \in L(l_2 \otimes \mathbb{C}^n) \otimes A$, and take an one-sided shift operator $S \in L(l_2)$, then set $W = J \otimes (I - p') + S \otimes p'$. Since $S^*S = J,\ J - SS^* = e_{11}$, where J is the unit of $L(l_2)$. It can be obtained by routine calculation,

$$W^*W = J \otimes (I - p') + S^*S \otimes p' = I,$$

and

$$I - WW^* = I - (J \otimes (I - p') + SS^* \otimes p') = e_{11} \otimes p'.$$

Thus W is an isometry. Let $V = (U \otimes 1)^*W(U \otimes 1)$, it follows from the above proof that $V \in L(l_2) \otimes A$ is an isometric module operator and $I - VV^* = P$.

Finally, we consider the general case. From the proof process of Lemma 2.4, there exists a unitary module operator $U \in L(H_A)$ such that $UPU^* \leq P_n$ whenever n is big enough. By the proof of the first part, there is an isometric module operator V' such that $I - V'(V')^* = UPU^*$. Take $V = U^*V'$, then by routine calculation, $I - VV^* = P$.

Remark 2.16 It is known from Lemma 2.5 that for any $P \in K(H \otimes A)$, there is an isometric operator such that $P = I - VV^*$.

The proof of Theorem 2.6. From Theorem 2.5, we see that the index mapping $Index : [T] \to Index(T)$ is a homomorphism. Now, to prove that $Index(,)$ is surjective. For this, suppose that E_1 and E_2 are algebraically finitely generated projective submodules in A^n. Set P and Q to be the projections from A^n onto E_1 and E_2, respectively. Then P and Q are in $M_n(A)(\subseteq K(H \otimes A))$. It follows from Lemma 2.5 that there exist isometric operators V and W in $L(H \otimes A)$ such that $I - VV^* = P$ and $I - WW^* = Q$. By routine calculation, we have

$$(WV^*)^*(WV^*) = VV^* = I - P,$$

and

$$WV^*(WV^*)^* = WW^* = I - Q.$$

Hence WV^* is an isometry. Since $\pi(W)$ and $\pi(V^*)$ are invertible, it is also $\pi(WV^*) = \pi(W)\pi(V^*)$. Therefore $WV^* \in \Phi(H \otimes A)$, and we have

$$Index(WV^*) = [ker\, WV^*] - [ker\, VW^*] =$$

$$= [E_1] - [E_2] \in K_0(A).$$

Thus $Index(,\ )$ is surjective.

We prove next that $Index(,\ )$ is injective. It suffices to prove that if $Index(T) = 0$ then $[T] = [I]$. Indeed, from Proposition 2.21, there exists $K \in K(H \otimes A)$ such that $V = T + K$ is invertible. Since $|V|$ is invertible, $0 \notin \sigma(|V|)$. Let $U = V|V|^{-1}$, then U is unitary by direct calculation. Then set $\psi(t) = U \cdot exp(t \log |V|)$ $(t \in [0, 1])$, in which each $\psi(t)$ is invertible, and $\psi(0) = u$, $\psi(1) = u|v| = v$ from above construction. Through simple calculation, we have $\psi^{-1}(t) = exp(-t \log |V|) \cdot U^*$. Since U is unitary, there is a continuous path consisted of unitary operators to connect U and I by applying Proposition 2.22. On the other hand, $\phi : t \to T + tK$ $(t \in [0, 1])$ connects T and V. Therefore from the above proof, we get $[T] = [I]$, where $[I]$ is the unit of $[\Phi(H \otimes A)]$.

Finally, to show that $[\Phi(H \otimes A)]$ is group, for this, it suffices to prove that every element in $[\Phi(H \otimes A)]$ is invertible. If $[T] \in [\Phi(H \otimes A)]$, then it is easy to check that $Index[T] = -Index[T^*]$. Hence

$$Index[TT^*] = Index[T][T^*]$$

$$= Index[T] + Index[T^*] = 0$$

$$= Index[I].$$

Since $Index(,)$ is injective, therefore $[T][T^*] = [I]$, so that $[T]^{-1} = [T^*]$.

2.4 Introduction to Modules Frames Theory

2.4.1 Existence of Module Frame and Reconstruction Formula

Frame theory on Hilbert space has important applications in wavelet analysis and signal processing. Module frame theory originates from the frame analysis in Hilbert space. Now module frame theory has been widely used in the study of Cuntz-Pimsner algebra, Jones index theory, noncommutative geometry, and operator valued free probability theory.

The spatial structure of Hilbert space can be determined by any set of orthogonal bases. Similarly, can a set of generators of Hilbert A-module determine its spatial structure (in the sense of unitary equivalence)? The answer is no; that is generally not true. Finding geometric invariants is one of the motivations to describe the framework of modules.

Let us briefly describe the frame concept of Hilbert space: suppose that H is a separable Hilbert. A sequence $\{x_i\}_{i=1}^{\infty} \subset H$ is called a frame of H if there exist positive constants C and D such that

$$C||x||^2 \leq \sum_{n=1}^{\infty} |< x, x_n >|^2 \leq D||x||^2 \ (x \in H).$$

It is emphasized here that if not specified, all C^*-algebras considered in this section have identity elements.

Similarly, the concept of frame in Hilbert A-module can be introduced as follows.

Definition 2.15 Let E be a Hilbert C^*-module over a unital C^*-algebra and J be a finite or countable index set. A sequence $\{x_j\}_{j \in J}$ of elements in E is said to be a standard frame if there are positive constants C and D such that

$$C < x, x > \leq \sum_{j \in J} < x, x_j >< x_j, x > \leq D < x, x >,$$

for all $x \in E$, in which the sum in the middle of the above inequality always converges in norm. Moreover, if $C = D = 1$ in the above inequality, then it is called standard normalized tight frame, or tight frame in short in the sequel.

Remark 2.17

(1) For the convenience of writing, if the index set J is countable, we may as well assume that it is the set of positive integers $\mathbb{N}$. If J is finite, It is regarded as 0 after a fixed item.

(2) If $\{x_i\}_{i=1}^{\infty}$ is a tight frame in a Hilbert A-module E, then it is easy to check that

$$\sum_{i=1}^{\infty} < x, x_i >< x_i, x >=< x, x >,$$

for all $x \in E$.

Example 2.4 Consider $H \otimes A$. Let $\{e_i\}_{i=1}^{\infty}$ be a standard orthogonal basis of H, then $\{e_i \otimes 1\}_{i=1}^{\infty}$ is a tight frame of E, where 1 is the unit of C^*-algebra A. In addition, each frame $\{x_i\}_{i=1}^{\infty}$ of H can induce the corresponding frame $\{x_i \otimes 1\}_{i=1}^{\infty}$. of $H \otimes A$.

In fact, since $H \otimes A \cong l_2(A)$, we regard $e_i \otimes 1$ and $(\underbrace{0, \cdots, 0}_{i-1}, 1, 0, \cdots)$ as the same in the sense of unitary equivalence for all $i \in \mathbb{N}$. Therefore, we might as well consider the existence of tight frame on $l_2(A)$. Set $f_i = (0, \cdots, 0, 1, 0, \cdots)$, $i = 1, 2, \cdots$. Since $< f_i, f_j >= 0, i \neq j$ and $< e_i, e_i >= 1, i, j = 1, 2, \cdots$,, we have for any $\tilde{a} = (a_1, a_2, \cdots, a_n, \cdots)$,

$$\tilde{a} = \sum_{i=1}^{\infty} f_i a_i = \sum_{i=1}^{\infty} f_i < f_i, \tilde{a} > .$$

Therefore

$$< \tilde{a}, \tilde{a} >=< \tilde{a}, \sum_{i=1}^{\infty} f_i < f_i, \tilde{a} >>$$

$$= \sum_{i=1}^{\infty} < \tilde{a}, f_i < f_i, \tilde{a} >>= \sum_{i=1}^{\infty} < \tilde{a}, f_i >< f_i, \tilde{a} > .$$

Thus $\{f_i\}_{i=1}^{\infty}$ is a tight frame of $l_2(A)$ by Definition 2.15.

Remark 2.18 The tight frame $\{f_i\}_{i=1}^{\infty}$ is also called the standard orthonormal basis of $l_2(A)$.

From Example 2.4 and Kasprov's stabilization theorem, one can obtain the following conclusion:

Theorem 2.7 *Every countably generated Hilbert A-module has a tight frame.*

Proof If E is a countably generated Hilbert A-module, then by Kasprov's stabilization theorem, we have

$$E \oplus l_2(A) \cong l_2(A).$$

Thus E can be regarded as a complemented submodule in $l_2(A)$ (in the sense of unitary equivalence), so there is a projection $P \in L(l_2(A))$ such that $P(l_2(A)) = E$. Let $\{f_i\}_{i=1}^{\infty}$ be a standard orthogonal basis of $l_2(A)$. We prove next that $\{P(f_i)\}_{i=1}^{\infty}$ is a tight frame of E. Since $x = \sum_{i=1}^{\infty} f_i < f_i, x >$ for all $x \in l_2(A)$, and $P(x) = x$ for all $x \in E$. It follows that

$$x = P(x) = P\left(\sum_{i=1}^{\infty} f_i < f_i, x >\right)$$

$$= \sum_{i=1}^{\infty} P(f_i) < f_i, x > \tag{2.1}$$

We substitute $< f_i, x >=< f_i, P(x) >=< P(f_i), x >$ into formula (2.1) and get $x = \sum_{i=1}^{\infty} P(f_i) < P(f_i), x >$. Therefore

$$< x, x >=< \lim_{n\to+\infty} \sum_{i=1}^{n} P(f_i) < P(f_i), x >, x >$$

$$= \lim_{n\to+\infty} \sum_{i=1}^{n} < x, P(f_i) >< P(f_i), x >,$$

which converges in norm topology on A. So that $\{P(f_i)\}_{i=1}^{\infty}$ is a tight frame of E by Definition 2.15.

Remark 2.19

(1) Every finitely generated projective module has tight frames, whose proof is similar to that in Theorem 2.7.
(2) Since

$$0 \le \sum_{i\neq j} < x_j, x_i >< x_i, x_j >=< x_j, x_j > - < x_j, x_j >^2,$$

for any fixed $j \in J$. So that $< x_j, x_j >\le 1, j = 1, 2, \cdots$.

From the proof of Theorem 2.7, the following equation is obtained:

$$x = \sum_{i=1}^{\infty} P(f_i) < P(f_i), x > \quad (x \in E).$$

This equation is called the reconstruction formula of tight frame $\{P(f_i)\}_{i=1}^{\infty}$. It is the generalization of Fourier expansion in Hilbert space. In fact, the reconstruction formula and the tight frame is mutually determined, that is the following conclusion.

Theorem 2.8 *Let E be a countably generated Hilbert A-module and sequence $\{x_i\}_{i=1}^{\infty}$ in E. Then $\{x_i\}_{i=1}^{\infty}$ is a tight frame of E if and only if the corresponding reconstruction formula holds.*

Proof If the reconstruction formula holds, the proof has been included in the proof process of Theorem 2.7.

Conversely, assume that $\{x_i\}_{i=1}^{\infty}$ is a tight frame of E. Define a A-linear mapping θ from to $l_2(A)$ given by

$$\theta(x) = \{< x_n, x >\}_{n=1}^{\infty} \quad (x \in E).$$

Since

$$< x, x >= \sum_{n=1}^{\infty} < x, x_n >< x_n, x >$$

$$= \sum_{n=1}^{\infty} < x_n, x >^* < x_n, x >,$$

for all $x \in E$, in which converges in norm in A. Thus $\{< x_n, x >\}_{n=1}^{\infty} \in l_2(A)$, which shows the mapping θ is well-defined. Since

$$< \theta(x), \theta(x) >= \sum_{n=1}^{\infty} < x_n, x >^* < x_n, x >$$

$$= \sum_{n=1}^{\infty} < x, x_n >< x_n, x >=< x, x >, \ for\ all\ x\ in\ E.$$

Hence θ is an isometric embedding operator from E to $l_2(A)$.

We prove next that θ has adjoint operator. For this, construct an operator on the norm dense subspace of $l_2(A)$ and then extend it to $l_2(A)$ by preserving the norm, that is the θ^* we want. Write $\widetilde{l_2(A)}$ for the set of all finite A-linear combinations of $\{f_i\}_{i=1}^{\infty}$, then $\widetilde{l_2(A)}$ is a norm dense subspace of $l_2(A)$, where $\{f_i\}_{i=1}^{\infty}$ is a tight frame in $l_2(A)$ as shown in Example 2.4. Define a A-linear mapping from $\{f_i\}_{i=1}^{\infty}$ to E given by

$$\theta^*(f_n a) = x_n a \ (a \in A, \ n \in \mathbb{N}).$$

Then for all $x \in E$ and $y \in \widetilde{l_2(A)}$, we have

$$< \theta(x), y >=< x, \theta^*(y) >,$$

since

$$< \theta(x), f_n >=< \sum_{i=1}^{\infty} f_i < x_n, x >, f_n >$$

$$=< x, x_n > \cdot \sum_{i=1}^{\infty} < f_i, f_n >=< x, x_n >$$

$$=< x, \theta^*(f_n) >,$$

for all $x \in E$ and $f_n \in \{f_i\}_{i=1}^{\infty}$. It follows from what we have proved and Corollary 1.1 that

$$||\theta^*(y)|| = \sup_{||x|| \leq 1} || < x, \theta^*(y) > ||$$

$$= \sup_{||x|| \leq 1} || < \theta(x), \ y > || \leq \sup_{||x|| \leq 1} ||\theta(x)|| ||y||$$

$$= ||y|| \ (y \in \widetilde{l_2(A)}).$$

Therefore θ^* can be extended preserving norm to the whole space $l_2(A)$, which is still denoted as θ^*. Moreover, combined with the continuity of the inner product, we obtain that

$$< \theta(x), \ y > = < x, \ \theta^*(y) > \ (x \in E, \ y \in l_2(A)),$$

which shows that θ^* is the adjoint of θ. Since $< \theta(x), \ \theta(x) > = < x, \ x >$ for all $x \in E$, then by the polarization identity, we have

$$< \theta(x), \ \theta(y) >$$

$$= \frac{1}{4}[< \theta(x) + \theta(y), \ \theta(x) + \theta(y) > - < \theta(x) - \theta(y), \ \theta(x) - \theta(y) >$$

$$+ i < \theta(x) + i\theta(y), \ \theta(x) + i\theta(y) > -i < \theta(x) - i\theta(y), \ \theta(x) - i\theta(y) >]$$

$$= \frac{1}{4}[< x + y, \ x + y > - < x - y, \ x - y >$$

$$+ i < x + iy, \ x + iy > -i < x - iy, \ x - iy >] = < x, \ y >,$$

for all $x, \ y \in E$. It shows that $< \theta^*\theta(x), \ y > = < x, \ y > \ (x, \ y \in E)$, namely, $\theta^*\theta = I$. Therefore

$$x = \theta^*\theta x = \theta^* \left(\sum_{n=1}^{\infty} f_n < x_n, \ x > \right)$$

$$= \sum_{n=1}^{\infty} \theta^*(f_n < x_n, \ x >)$$

$$= \sum_{n=1}^{\infty} \theta^*(f_n) < x_n, \ x >$$

$$= \sum_{n=1}^{\infty} x_n < x_n, \ x > .$$

Thus, the reconstruction formula holds.

Remark 2.20

(1) Since $\theta\theta^*\theta = \theta$, $R(\theta\theta^*) = R(\theta)$. Since $R(I - \theta\theta^*) = ker\theta^*$, and $x = \theta\theta^*(x) + (I - \theta\theta^*)(x)$ for all $x \in l_2(A)$. Then $l_2(A) = \theta(E) \oplus ker\theta^*$.
    The above equation shows that the range of θ is a complemented submodule in $l_2(A)$, and $\theta : E \to \theta(E)$ is a unitary operator. Therefore θ is called frame transformation operator.
(2) From the reconstruction formula, we can see that the tight frame of E must be a generator set. But the opposite is not true.

Example 2.5 Let A be $C_0((0, 1])$ and E be $l_2(A)$. According to the Stone-Weierstrass approximation theorem, function $f(t) = t$ $(t \in (0, 1])$ is the topological generator of $C_0((0, 1])$. Let $f_i = \underbrace{(0, 0, \cdots, f}_{i}, 0, \cdots)$. Then $\{f_i\}_{i\in\mathbb{N}}$ is the set of generators of E, but $\{f_i\}_{i\in\mathbb{N}}$ is not a frame of E. In fact, assume that $\{f_i\}_{i\in\mathbb{N}}$ is a frame of E, then there exist positive integers C and D such that

$$C < x, x > \leq \sum_{j=1}^{\infty} < x, f_j >< f_j, x > \leq D < x, x > \quad (x \in E).$$

Let $x = f_1$ in the above equation; we get $C < f_1, f_1 > \leq < f_1, f_1 >^2$, that is, $Ct^2 \leq t^4$ $(t \in (0, 1])$,; it follows that $C \leq t^2$ $(t \in [0, 1])$; this is a contradiction.

But for algebraically finitely generated Hilbert A-modules, we have the following conclusion:

Theorem 2.9 *The algebraic generator set of an algebraically finitely generated Hilbert A-module is a standard frame. Thus, the standard frame is identity with its algebraic generator set.*

Proof Suppose that $\{x_i\}_{i=1}^n$ is an algebraic generated set of E. Define a A-linear mapping θ from E to A^n given by

$$\begin{aligned}\theta(x) &= (< x_1, x >, < x_2, x >, \cdots, < x_n, x >)\\ &= \sum_{i=1}^{n} f_i < x_i, x > = \sum_{i=1}^{n} \theta_{f_i, x_i}(x) \ (x \in E),\end{aligned}$$

where $\{f_i\}_{i=1}^n$ is a standard frame of A^n. Also, define $\theta^* : A^n \to E$ by the following way

$$\theta^*(a) = \sum_{i=1}^{n} x_i a_i = \sum_{i=1}^{n} \theta_{x_i, f_i}(a),$$

for all $a = \sum_{i=1}^{n} f_i a_i \in A^n$, where $a_1,\ a_2,\ \cdots,\ a_n \in A$. It is easy to check that

$$< \theta(x),\ a >= \sum_{i=1}^{n} < x,\ x_i a_i >=< x,\ \theta^*(a) > \quad (x \in E,\ a \in A^n),$$

which shows that θ^* is the adjoint of θ. Since E is generated linearly by $\{x_i\}_{i=1}^{n}$, then it follows from the sense of θ^* that $R(\theta^*) = E$. Thus by Proposition 1.8 $R(\theta)$ is a closed submodule of A^n also, and $A^n = ker(\theta^*) \oplus R(\theta)$. Let $\theta^* = V|\theta^*|$ be the polar decomposition of θ^*. Then we have $ker V = ker\theta^*$ and $V^*(E) = (ker V)^{\perp} = R(\theta)$. In what follows, we shall prove that $\{V(f_i)\}_{i=1}^{n}$ is a tight frame of E. Since $R(V) = R(\theta^*) = E$, so that if $x \in E$, then there exists $a \in A^n$ such that $Va = x$. Therefore

$$\sum_{i=1}^{n} < x,\ V(f_i) >< V(f_i),\ x >$$

$$= \sum_{i=1}^{n} < V(a),\ V(f_i) >< V(f_i),\ V(a) >$$

$$= \sum_{i=1}^{n} < V^*V(a),\ f_i >< f_i,\ V^*V(a) >$$

$$=< V^*V(a),\ V^*V(a) >$$

$$=< VV^*V(a),\ V(a) >$$

$$< V(a),\ V(a) >$$

$$=< x,\ x >,$$

which shows that $\{V(f_i)\}_{i=1}^{n}$ is a tight frame of E. Note that $\theta^* = \theta^*V^*V$,; therefore combined with the Definition of θ^*, it is concluded that $x_i = \theta^*(f_i) = \theta^*V^*V(f_i) = \theta^*V^*(V(f_i)), i = 1, 2, \cdots, n$. It follows that

$$\sum_{i=1}^{n} < x,\ x_i >< x_i,\ x >$$

$$= \sum_{i=1}^{n} < x,\ \theta^*V^*(V(f_i)) >< \theta^*V^*(V(f_i)),\ x >$$

$$= \sum_{i=1}^{n} < V\theta x,\ V(f_i) >< V(f_i),\ V\theta x >$$

$$=< V\theta x,\ V\theta x >$$

$$=< (\theta^* V^*)^* x,\ (\theta^* V^*)^* x > \tag{2.2}$$

Since $\theta^* V^*$ is invertible in $L(E)$, then from Proposition 1.3, it implies that

$$||(\theta^* V^*)^{-1}||^{-2} < x,\ x > \leq \sum_{i=1}^{n} < x,\ x_i >< x_i,\ x > \leq ||\theta^* V^*||^2 < x,\ x >,$$

which shows that $\{x_i\}_{i=1}^n$ is a standard frame of E.

From the proof of Theorem 2.9, we can see that the standard frame $\{x_i\}_{i=1}^n$ of algebraically finitely generated Hilbert A-module is actually the image set of tight frame $\{V(f_i)\}_{i=1}^n$ under the action of adjointable and invertible module operator $\theta^* V^*$. This phenomenon is not accidental. It reveals the internal relationship between the standard frame and the tight frame.

Theorem 2.10 *If $\{x_j\}_{j\in\mathbb{N}}$ is a standard frame of countable generated or finite generated Hilbert A-module E, then $\{(\theta^*\theta)^{-1/2} x_j\}_{j\in\mathbb{N}}$ is a tight frame of E. Conversely, for any tight frame $\{y_j\}_{j\in\mathbb{N}}$ of E, then $\{(\theta^*\theta)^{1/2} y_j\}_{j\in\mathbb{N}}$ is a standard frame of E, where θ is the frame transformation operator of E.*

Proof Let $\{f_i\}_{i=1}^\infty$ be the standard orthogonal basis of $l_2(A)$. Since $\{x_i\}$ is a standard frame of E, from the frame transformation

$$\theta : E \to l_2(A),\ \theta(x) = \{< x_j,\ x >\}_{j=1}^\infty = \sum_{j=1}^{\infty} f_j < x_j,\ x > \ \ (x \in E),$$

there exist positive constants C and D such that

$$C < x,\ x > \leq < \theta(x),\ \theta(x) > \leq D < x,\ x >,$$

for all $x \in E$, which shows that θ is injective and has closed range,

$$||\theta|| = \sup\{||\theta x|| : ||x|| \leq 1\} \leq \sqrt{D}.$$

Define a A-linear mapping θ^* from $\widetilde{l_2(A)}$ to E given by the following way: $\theta^*(f_n) = x_n$, where $\widetilde{l_2(A)}$ is the set of all finite A-linear combinations of $\{f_i\}_{i=1}^\infty$ which is dense in $l_2(A)$. Since

$$< \theta(x),\ f_n >= \sum_{j\in J} < x,\ x_j >< f_j,\ f_n >$$

$$=< x,\ x_n >=< x, \theta^*(f_n) > .$$

Therefore on $\widetilde{l_2(A)}$, we have

$$||\theta^*|| = \sup\{|| < \theta^* y,\ x > || : y \in \widetilde{l_2(A)},\ x \in E,\ ||y|| \le 1,\ ||x|| \le 1\}$$

$$= \sup\{|| < y, \theta x > ||\} \le ||\theta|| \le \sqrt{D}.$$

Thus θ^* can be extended preserving norm to $l_2(A)$ and becomes the adjoint of θ. Moreover, because θ is injective and has closed range, so θ^* is surjective and $l_2(A) = \theta(E) \oplus ker\theta^*$. Thus θ^* is invertible when restricted to $\theta(E)$, and therefore $\theta^*\theta$ is an invertible element in $L(E)$. Set $y_j = (\theta^*\theta)^{-1/2}(x_j),\ j \in \mathbb{N}$.

We prove next that $\{y_j\}_{j\in\mathbb{N}}$ is a tight frame of E. Since for any $x \in E$, there exists $y \in E$ such that $(\theta^*\theta)^{1/2}(y) = x$. It follows that

$$< x, x >=< (\theta^*\theta)^{1/2}(y),\ (\theta^*\theta)^{1/2}(y) >$$

$$=< \theta(y),\ \theta(y) >= \sum_{j\in J} < y,\ x_j >< x_j, y >$$

$$= \sum_{j\in\mathbb{N}} < (\theta^*\theta)^{-1/2}(x),\ x_j >< x_j,\ (\theta^*\theta)^{-1/2}(x) >$$

$$= \sum_{j\in\mathbb{N}} < x,\ (\theta^*\theta)^{-1/2}(x_j) >< (\theta^*\theta)^{-1/2}(x_j),\ x >$$

$$= \sum_{j\in\mathbb{N}} < x,\ y_j >< y_j,\ x > .$$

The above proof shows that $\{y_j\}_{j\in\mathbb{N}}$ is a tight frame of E.

Conversely, if $\{y_j\}_{j\in\mathbb{N}}$ is a tight frame of E, then it is easy to check that $\{(\theta^*\theta)^{1/2}(y_j)\}_{j\in\mathbb{N}}$ is a standard frame of E.

Remark 2.21 In fact, for any invertible module operator $T \in L(E)$, if $\{x_j\}_{j\in\mathbb{N}}$ is a tight frame of E, then it can be proved that $\{Tx_j\}_{j\in\mathbb{N}}$ is a standard frame of E.

According to Theorem 2.8, tight frame has reconstruction formula. Through Theorem 2.10, it is naturally hoped that similar formula will hold for the standard frame. Indeed, if $\{x_j\}_{j\in\mathbb{N}}$ is a standard frame of a countable generated or finite generated Hilbert A-module E, denoted by θ the frame transformation of E. Let $T = (\theta^*\theta)^{-1/2}$, then $\{T(x_j)\}_{j\in\mathbb{N}}$ is a tight frame of E by Theorem 2.10. Therefore

$$\sum_{n=1}^{\infty} x_n < T^*T(x_n),\ x >= \sum_{n=1}^{\infty} x_n < T(x_n),\ T(x) >$$

$$= \sum_{n=1}^{\infty} T^{-1}(T(x_n)) < T(x_n),\ T(x) >$$

$$= T^{-1}\left(\sum_{n=1}^{\infty} T(x_n) < T(x_n),\ T(x) >\right)$$

$$= T^{-1}(T(x)) = x,$$

from which the reconstruction formula of the standard frame is obtained as follows:

$$x = \sum_{n=1}^{\infty} x_n < (\theta^*\theta)^{-1}(x_n), x > \quad (x \in E).$$

$(\theta^*\theta)^{-1}$ is called frame operator. In particular, if $\{x_n\}$ is itself a tight frame, then

$$(\theta^*\theta)^{-1} = \theta^*\theta = I,$$

where I is the identity module operator. Furthermore, from the above reconstruction formula, it can be concluded that

$$(\theta^*\theta)^{-1}(x) = (\theta^*\theta)^{-1}\left(\sum_{n=1}^{\infty} x_n < (\theta^*\theta)^{-1}(x_n),\ x >\right)$$

$$= \sum_{n=1}^{\infty} (\theta^*\theta)^{-1}(x_n) < x_n, (\theta^*\theta)^{-1}x > .$$

Let $y = (\theta^*\theta)^{-1}(x)$, then

$$y = \sum_{n=1}^{\infty} (\theta^*\theta)^{-1}(x_n) < x_n,\ y > .$$

Thus $\{(\theta^*\theta)^{-1}(x_n)\}$ is called canonical dual frame of $\{x_n\}$.

2.4.2 Module Frames and Unitary Equivalence Between Hilbert C*-Modules

Theorem 2.11 *Let M and N be Hilbert C^*-modules over a C^*-algebra A, and let $\{x_j\}$ and $\{y_j\}$ be tight frames of M and N, respectively. Then the following conclusions are equivalent:*

(1) M and N are unitary equivalence which keeping frame, that is, there exists a unitary module operator $U : M \to N$ such that $Ux_n = y_n$;
(2) Their frame transformation θ_M and θ_N have the same ranges in $l_2(A)$, respectively.

Proof (1) $\Rightarrow$ (2) Suppose that there exists a unitary $U : M \to N$ such that $Ux_n = y_n$. Hence for any $y \in N$, there is a $x \in M$ such that $Ux = y$. Note that frame transformations

$$\theta_M : M \to l_2(A,\ \theta_M(x)) = \{< x_n,\ x >\}_{n=1}^{\infty}\ (x \in M),$$

and

$$\theta_N : N \to l_2(A,\ \theta_N(x)) = \{< x_n,\ x >\}_{n=1}^{\infty}\ (x \in N).$$

Therefore

$$\theta_N(y) = \theta_N(U(x)) = \{< Ux_n,\ Ux >\}_{n=1}^{\infty}$$

$$= \{< x_n,\ x >\}_{n=1}^{\infty} = \theta_M(x),$$

which shows that $R(\theta_N) \subset R(\theta_M)$. Similarly, $R(\theta_M) \subset R(\theta_N)$. Thus $R(\theta_M) = R(\theta_N)$.

(2) $\Rightarrow$ (1) Note that $\theta_M : M \to \theta_M(M)$ and $\theta_N : N \to \theta_N(N)$ are unitary. From the assumption $\theta_M(M) = \theta_N(N)$, we get that $\theta_N^*\theta_M : M \to \theta_M(M) = \theta_N(N) \to N$ is unitary. Concretely,

$$< \theta_N^*\theta_M(x),\ \theta_N^*\theta_M(x) >=< \theta_N\theta_N^*\theta_M(x),\ \theta_M(x) >$$

$$=< \theta_M(x),\ \theta_M(x) >=< x,\ x > \quad (x \in M).$$

Remark 2.22 Because $l_2(A) = \theta_M(M) \oplus ker\theta_M^*$ and $l_2(A) = \theta_N(N) \oplus ker\theta_N^*$, so $\theta_M(M) = \theta_N(N) \Leftrightarrow ker\theta_M^* = ker\theta_N^*$. Since for any $x \in l_2(A)$, we have $x = \sum_{n=1}^{\infty} f_n < f_n,\ x >$. Hence

$$\theta_M^*(x) = \sum_{n=1}^{\infty} \theta_M^*(f_n) < f_n, x >= \sum_{n=1}^{\infty} x_n < f_n,\ x >,$$

and

$$\theta_N^*(x) = \sum_{n=1}^{\infty} \theta_N^*(f_n) < f_n, x >= \sum_{n=1}^{\infty} y_n < f_n, x > .$$

for all $x \in l_2(A)$. Therefore

$$ker\theta_M^* = \{x \in l_2(A) : \sum_{n=1}^{\infty} x_n < f_n, x >= 0\},$$

and

$$ker\theta_N^* = \{x \in l_2(A) : \sum_{n=1}^{\infty} y_n < f_n, x >= 0\}.$$

It is shown that the elements with the same A-linear component in M and N are both zero or nonzero, which is also a necessary and sufficient condition for unitary equivalence keeping frame between M and N.

In particular, for finitely generated projective modules, there is a more specific unitary equivalence theorem as follows:

Theorem 2.12 *Let E and F be finite generated projective module. Then the following conclusions are equivalent:*

(1) E and F are unitary equivalence;
(2) There exist respectively tight frames $\{x_1, \cdots, x_k\}$ and $\{y_1, \cdots, y_l\}$ of E and F such that $k = l$ and $< x_i, x_j >=< y_i, y_j >$.

The proof is left to the reader as an exercise.

2.4.3 Unitary Equivalence of Closed Submodules and Stable Isomorphism Between Hereditary C^*- Subalgebras

Stable isomorphism between hereditary C^*-subalgebras plays an important role in the K-theory of C^*-algebras (see [3, 11, 21]). L.G. Brown proved in [3] that for σ-unital C^*-algebra A, its full hereditary C^*-subalgebras are stable isomorphic to A. Moreover, if $||p-q|| < 1$ then $Her(p)$ is isomorphic to $Her(q)$, where p and q are open projections in A^{**}, and $Her(p)$ and $Her(q)$ are hereditary C^*-subalgebras generated by p and q in A, respectively. The concepts of open and closed projections can be referred to [18].

Let us consider the inverse problem of the above conclusion: if $Her(p)$ is isomorphic to $Her(q)$, then under what conditions can we get $p \sim q$? In the following, we shall use module frame theory to study the above problem.

Theorem 2.13 *Let E be a Hilbert A-module. Then there is a bijection between the set of all closed submodules in E and the set of all hereditary C^*-subalgebras of $K(E)$. Concretely, there is a bijection given by*

$$\Phi : \mathbb{E} \to \mathbb{K},\ E_0 \to B_{E_0},$$

where $B_{E_0} = \{T \in K(E) : TE \subset E_0,\ T^*E \subset E_0\}$ is a hereditary C^*-subalgebra of $K(E)$ associated with E_0, $\mathbb{E}$ is the set of all closed submodules in E, and $\mathbb{K}$ is the set of all hereditary C^*-subalgebras of $K(E)$.

We first need a lemma.

Lemma 2.6 *Let E be a Hilbert A-module and a closed submodule E_0 of E. Set $B_{E_0} = \{T \in K(E) :\ TE \subset E_0,\ T^*E \subset E_0\}$, then as C^*-algebras, there exists isometric $*$ isomorphism between $K(E_0)$ and B_{E_0}.*

Proof It is easy to check that the above B_{E_0} is a hereditary C^*-subalgebra of $K(E)$. We write $K_0(E_0)$ for the set of all the finite linear combinations of the form $\theta_{x,\,y}$ $(x,\ y \in E_0)$, then $K_0(E_0)$ is dense in $K(E_0)$, from which it implies that $K_0(E_0) \subset K(E)$, since $\theta_{x,\,y} \in K(E)$ for all $x,\ y \in E_0$. In what follows, the key is to prove that

$$||\sum_{i=1}^{n} \theta_{x_i,\,y_i}|E_0|| = ||\sum_{i=1}^{n} \theta_{x_i,\,y_i}||,$$

for all $x_i,\ y_i \in E_0$ $(i = 1,\ 2,\ \cdots,\ n)$. When $x \in E_0$, from Proposition 1.4, we have

$$||\theta_{x,\,x}|| = ||<x,\ x>|| = ||\theta_{x,\,x}|E_0||.$$

Hence

$$||\theta_{x,\,y}|| = ||\theta^*_{x,\,y}\theta_{x,\,y}||^{1/2}|| = ||\theta_{y<x,\,x>^{1/2},\ y<x,\,x>^{1/2}}||$$

$$= ||<x,\ x>^{1/2}<y,\ y>^{1/2}|| = ||\theta_{x,\,y}|_{E_0}||,$$

for all $x,\ \ y\ \in\ E_0$. Since $E_0 \otimes \mathbb{C}^n$ is a Hilbert $M_n(A)$-module, $\theta_{x,\,y} = (\sum_{i=1}^{n} \theta_{x_i,\,y_i}) \otimes I_n$ when $x = (x_1,\ \cdots,\ x_n)$ and $y = (y_1,\ \cdots,\ y_n) \in E_0^n = E_0 \otimes \mathbb{C}^n$, where I_n is the unit of $M_n(\mathbf{C})$. It results from Corollary 1.2 that

$$||\sum_{i=1}^{n} \theta_{x_i,\,y_i}|| = ||\theta_{x,\,y}|| = ||<x,\ x>^{1/2}_{M_n(A)}<y,\ y>^{1/2}_{M_n(A)}|| = ||\sum_{i=1}^{n} \theta_{x_i,\,y_i}|E_0|| \tag{2.3}$$

Further, if $T \in K(E_0)$, then there is a Cauchy sequence $(T_n) \subset K_0(E_0)$ such that

$$||T_n - T|| \to 0(n \to \infty).$$

From the above Equation (2.3), we see that (T_n) is also the Cauchy sequence in $K(E)$. Thus there exists an element $\tilde{T} \in K(E)$ such that

$$||T_n - \tilde{T}|| \to 0(n \to \infty).$$

The above proof shows that T has an unique extension $\tilde{T}$ in $K(E)$ and $||T|| = ||\tilde{T}||$. Because $T_n E \subset E_0,\ T_n^* E \subset E_0$, so $\tilde{T} E \subset E_0,\ \tilde{T}^* E \subset E_0$. Hence $\tilde{T} \in B_{E_0}$.

We now define a linear mapping Ψ from $K(E_0)$ to B_{E_0} given by

$$\Psi : K(E_0) \to B_{E_0}, T \to \tilde{T},\ (T \in K(E_0)).$$

It follows from what we have proved and by routine calculation that Ψ is an isometric $*$ isomorphism from $K(E_0)$ onto B_{E_0}.

The proof of Theorem 2.13. Define a linear mapping Φ from $\mathbb{E}$ to $\mathbb{B}$ given by the following way

$$\Phi : \mathbb{E} \to \mathbb{B},\ E_0 \to B_{E_0}\ (E_0 \in \mathbb{E}).$$

Then Φ is well-defined from Lemma 2.6. We first show that Φ is injective. Equivalently, we need to prove that if $E_1,\ E_2 \in \mathbb{E}$ and $B_{E_1} = B_{E_2}$, then $E_1 = E_2$. From the sense of B_{E_1}, we have $[B_{E_1} E] \subset E_1(write[\ ]$ for the norm closure). It follows from Lemma 2.6 that $B_{E_1} \cong K(E_1)$ and $B_{E_1}|E_1 = K(E_1)$. Observe that $[K(E_1)E_1] = E_1$, hence $[B_{E_1}]E \supset [B_{E_1} E_1] = E_1$. Thus $[B_{E_1} E] = E_1$. Similarly, $[B_{E_2} E] = E_2$. Hence $E_1 = [B_{E_1} E] = [B_{E_2} E] = E_2$ is resulted from $B_{E_1} = B_{E_2}$.

We prove next that Φ is surjective. For this, we want to prove that every hereditary C^*-subalgebra B of $K(E)$ can be written in the form $B = B_{E_B}$, where $E_B = [BE]$. It is evident that $B \subset B_{E_B}$. In the following, the key is to prove $B_{E_B} \subset B$. Let H_1 be the set of all finite linear combinations of the form $\theta_{x,\ y}\ (x,\ y \in E_B)$; then it is easy to check that $[H_1] = K(E_B)$. Also, let H_0 be the set of all finite linear combinations of the form $(\theta_{Tx,\ Sy}\ (T, S \in B,\ x,\ y \in E)$. Then $[H_0] \subset B$, since $\theta_{Tx,\ Sy} = T\theta_{x,\ y}S^* \in B$ for all $S,\ T \in B,\ x,\ y \in E$. Observe that $H_1 \subset [H_0]$ from the constructions of them; it follows from the above that $K(E_B) = [H_1] \subset B \subset B_{E_B}$. From Lemma 2.6, there exists an surjective $*$ homomorphism Ψ from $K(E_B)$ onto B_{E_B} such that $B_{E_B}|E_B = K(E_B)$. Since $K(E_B) \subset B|E_B \subset B_{E_B}|E_B$ by formula (2.4), $B|E_B = K(E_B)$. Therefore from Lemma 2.6 again we see that $\Psi(K(E_B)) = B$, so $B = B_{E_B}$.

From Theorem 2.13, the following corollary is immediately obtained:

Corollary 2.5 (see [24]) *If E is a Hilbert A-module. Then every closed submodule F of E is complemented in E, if and only if every hereditary C^*-subalgebra B of*

$K(E)$ is of the form $B = pK(E)p$, where p is a projection in $L(E)$ associated with F.

In Hilbert space theory, two separable Hilbert spaces H_1 and H_2 are unitary equivalent if and only if $K(H_1)$ and $K(H_2)$ are $*$ isomorphic. Similarly, for Hilbert A-modules E and F, it is easily proved that if E and F are unitary equivalent, then $K(E)$ and $K(F)$ are $*$ isomorphic. But the opposite is not true in general.

For example, let A be a hyperfinite II_1-factor, and let $E_1 = A,\ E_2 = A \oplus A$, then as Hilbert A-modules, we have $A \cong K(E_1) \cong K(E_2)$. But it can be proved that E_1 and E_2 are not unitary equivalent.

However, the following conclusion is obtained by using the module frame theory.

Theorem 2.14 *Let E and F be finite generated projective modules over a unital C^*-algebra A, that is, there exists free module A^{N_0} such that E and F are complemented submodules in A^{N_0}, respectively, where N_0 is a fixed positive integer. Then E and F are unitary equivalent if and only if the following two conditions hold:*

(1) There is a $*$ isomorphism Φ from $K(E)$ onto $K(F)$;
(2) For any given positive integer m $(1 \leq m \leq N_0)$ and any sequence $x^m = (x_i)_{i=1}^m \subset E^m$, there exists sequence $x'^m = (x_i')_{i=1}^m \subset F^m$ such that

$$\Phi^{(m)}(\theta_{x^m a,\, x^m}) = \theta_{x'^m a,\, x'^m},$$

for all $a \in M_m(A)$, in which $\Phi^{(m)} = \Phi \otimes I_m$ is a $*$ isomorphism from $K(E) \otimes \mathbb{C}^m$ to $K(F) \otimes \mathbb{C}^m$ induced by Φ, I_m is the identity mapping on $\mathbb{C}^m$.

Proof Necessity: If there exists a unitary $U : E \mapsto E_2$, then it is easy to check that the linear mapping Φ from $K(E_1)$ onto $K(E_2)$ defined by $\Phi(T) = U \circ T \circ U^*$ is a $*$ isomorphism, and the condition (2) in Theorem 2.14 is also satisfied.

Conversely, if there exists a $*$ isomorphism Φ from $K(E_1)$ onto $K(E_2)$ satisfying also the condition (2), then it follows easily that for every x in E, there exists a $x' \in F$ associated with x, such that

$$\Phi(\theta_{xa,\, xa}) = \theta_{x'a,\, x'a},$$

for each fixed $a \in A$. Therefore

$$\|<xa,\ xa>\| = \|\theta_{xa,\, xa}\|$$

$$= \|\Phi(\theta_{xa,\, xa})\| = \|\theta_{x'a,\, x'a}\|$$

$$= \|<x'a,\ x'a>\|.$$

Hence $\|a^* <x,\ x> a\| = \|a^* <x',\ x'> a\|$, from which by using the basic properties of C^*-algebra, we have $\|<x,\ x>^{1/2} a\|^2 = \|<x',\ x'>^{1/2} a\|^2$ for any fixed $a \in A$. Thus, by Lemma 1.1, we get $< x, x >^{1/2} = < x', x' >^{1/2}$ and is therefore $< x, x > = < x', x' >$. In what follows, let $\{e_i\}_{i=1}^{T}$ be a standard frame of E, where $T(\leq N_0)$ is a positive integer. Then

$$\sum_{i=1}^{T} \theta_{e_i, e_i} = 1.$$

Applying the above proof, there exists a sequence $\{f_i\}_{i=1}^{T}$ included in F such that

$$\Phi(\theta_{e_i, e_i}) = \theta_{f_i, f_i}, \ i = 1, 2, \cdots, T.$$

It implies that

$$\sum_{i=1}^{T} \theta_{f_i, f_i} = \Phi\left(\sum_{i=1}^{T} \theta_{e_i, e_i}\right) = 1,$$

which shows that $\{f_i\}_{i=1}^{T}$ is a standard frame of F. Using the assumption (2), for any fixed $a \in M_T(A)$, we have

$$\Phi^{(T)}(\theta_{(e_i)_{i=1}^{T} a, (e_i)_{i=1}^{T}}) = \theta_{(f_i)_{i=1}^{T} a, (f_i)_{i=1}^{T}}.$$

It follows that

$$\begin{pmatrix} < e_1, f_1 >, & < e_1, e_2 >, & \ldots < e_1, e_T > \\ < e_2, e_1 >, & < e_2, e_2 >, & \ldots < e_2, e_T > \\ \vdots & & \vdots \\ < e_T, e_1 >, & < e_T, e_2 >, & \ldots < e_T, e_T > \end{pmatrix}$$

$$= \begin{pmatrix} < f_1, f_1 >, & < f_1, f_2 >, & \ldots < f_1, f_T > \\ < f_2, f_1 >, & < f_2, f_2 >, & \ldots < f_2, f_T > \\ \vdots & & \vdots \\ < f_T, f_1 >, & < f_T, f_2 >, & \ldots < f_T, f_T > \end{pmatrix}.$$

Hence $< e_i, e_j > = < f_i, f_j >, \ i, j = 1, 2, \cdots T$. In the end, define a A-linear mapping U from E to F given by

$$U(e_i) = f_i \ (i = 1, 2, \cdots, T).$$

To show that U is unitary, it suffices to prove that U is an isometry by using Theorem 1.2. Since for all $x \in E$, we have

$$x = \sum_{i=1}^{T} e_i < e_i, x >, \quad U(x) = \sum_{i=1}^{T} f_i < e_i, x > .$$

Therefore

$$< U(x), U(x) > = < \sum_{i=1}^{T} f_i < e_i, x >, \sum_{j=1}^{T} f_j < e_j, x >>$$

$$= \sum_{i=1}^{T} \sum_{j=1}^{T} < f_i < e_i, x >, f_j < e_j, x >>$$

$$= \sum_{i=1}^{T} \sum_{j=1}^{T} < x, e_i >< f_i, f_j >< e_j, x >$$

$$= \sum_{i=1}^{T} \sum_{j=1}^{T} < x, e_i >< e_i, e_j >< e_j, x >$$

$$= \sum_{i=1}^{T} < x, e_i >< e_i, \sum_{j=1}^{T} e_j < e_j, x >>$$

$$= \sum_{i=1}^{T} < x, e_i >< e_i, x > = < x, x > .$$

Thus U is an isometry.

At the end of this subsection, we shall give the relations among the unitary equivalence between closed submodules, the $*$ isomorphism of the corresponding hereditary C^*-subalgebras, and the equivalence of the corresponding open projections based on $H \otimes A$.

Theorem 2.15 *Let A be a unital C^*-algebra and H a separable Hilbert space, and let E_1 and E_2 be finite generated projective submodules in $H \otimes A$, p_1 and p_2 (in $L(H \otimes A)$) are the projections from $H \otimes A$ onto E_1 and E_2, respectively.Then the following conditions are equivalent:*

(1) E_1 and E_2 are unitary equivalent;
(2) There exists a surjective $*$ isomorphism $\Phi : K(E_1) \to K(E_2)$, and Φ preserves the finite rank module operator up to N_0 (in the sense of Theorem 2.14*);*

(3) $p_1 \sim p_2$ *(Murray-von Neumann equivalence, that is, there exists partial isometry* $v \in L(l^2(A))$ *such that* $v^*v = p_1,\ vv^* = p_2$*).*

Proof (1) $\Leftrightarrow$ (2) is obvious from Theorem 2.14. Now we prove that (1) $\Leftrightarrow$ (3) : Because $E = p_1(H \otimes A)$ and $F = p_2(H \otimes A)$, so that if there exists a unitary U in $L(E,\ F)$, then U can be extended to a partially isometric module operator $\tilde{U}$ in $L(H \otimes A)$, such that

$$\tilde{U}^*\tilde{U} = p_1,\ \tilde{U}\tilde{U}^* = p_2,$$

which shows that $p \sim q$

Conversely, if $p_1 \sim p_2$, then there exists a partially isometry V in $L(H \otimes A)$ such that $V^*V = p$ and $VV^* = q$. Thus, the restriction $V|E$ is an isometric module operator from E to F, and therefore $V|E \in L(E,\ F)$, since E and F are complemented submodules, which follow from Theorem 1.2 that $V|E$ is unitary from E onto F.

References

1. Blackadar, B., Handelman, D.: Dimension functions and traces on C^*-algebras. J. Funct. Anal. **45**, 297–340 (1982)
2. Bratteli, O., Robinson, D.W.: Operator Algebras and Quantum Statistical Mechanics II. Equilibrium States Models in Quantum Statistical Mechanics. Springer, New York/Berlin/Heidelberg (1981)
3. Brown, L.G.: Stable isomorphism of hereditary subalgebras of C^*-algebras. Pac. J. Math. **71**(2), 335–348 (1977)
4. Brown, L., Green, P., Rieffel, M.: Stable isomorphism and strong Morita equivalence of C^*-algebras. Pac. J. Math. **71**, 349–363 (1977)
5. Carlen, E.A., Lieb, E.H.: Optimal hypercontractivity for Fermi fields and related noncommutative integration inequalities. Commun. Math. Phys. **155**, 27–46 (1993)
6. Cuntz, J.: Simple C^*-algebras generated by isometries. Commun. Math. Phys. **57**, 173–185 (1977)
7. Cuntz, J., Higson, N.: Kuiper's theorem for Hilbert modules. In: Jorgensen, P.E.T., Muhly, P.S. (eds.) Operator Algebras and Mathematical Physics. Contemporary Mathematics, vol. 62, pp. 429–434 (1987). AMS
8. Dixmier, J.: C^*-Algebras. North-Holland, Amsterdam (1982)
9. Douglas, R.G.: Banach Algebras Techniques in Operator Theory. Academic, New York (1972)
10. Frank, M.: Geometrical aspects of Hilbert C^*-modules. Positivity **3**, 215–243 (1999)
11. Huaxin Lin: Equivalent open projections and corresponding hereditary C^*-subalgebras. J. Lond. Math. Soc. (2) **41**, 295–301 (1990)
12. Kajiwara, T., Pinzari, C., Watatani, Y.: Ideal structure and simplicity of the C^*-algebras generated by Hilbert bimodules. J. Funct. Anal. **159**, 295–322 (1998)
13. Kasparov, G.G.: The operator K-functor and extensions of C^*-algebras. Izv. Akad. Nauk SSSR. Ser. Mat. **44**(3), 571–636 (1980)
14. Li Bengren, Banach Algebras. China Science Press, Beijing (1992)
15. Mingo, J.A.: K-theory and multipliers of stable C^*-algebras. Trans. Am. Math. Soc. **299**, 397–411 (1987)

16. Muhly, P.S., Solel, B.: On the Morita equivalence of tensor alhgebras. Proc. Lond. Math. Soc. **81**(3), 113–168 (2000)
17. Murphy, G.J.: C^*-Algebras and Operator Theory. Academic, London (1990)
18. Pedersen, G.K.: C^*-Algebras and Their Automorphism Groups. Academic, London (1979)
19. Pimsner, M.V.: A class of C^*-algebras generalizing both Cuntz-Krieger algebras and crossed products by $\mathbb{Z}$. Fields Inst. Commun. **12**, 189–212 (1997)
20. Rieffel, M.A.: Induced representations of C^*-algebras. Adv. Math. **13**, 176–257 (1974)
21. Shuang Zhang, Certain C^*-algebras with real rank zero and the internal structure of their corona and multiplier algebras, Part I. Pac. J. Math. **115**(1), 169–179 (1992)
22. Speicher, R.: Combinatorial Theory of the Free Product with Amalgamation and Operator-Valued Free Probability Theory. Memoirs AMS, vol. 627. AMS, Providence, RI (1998)
23. Wegge-Olsen, N.E.: K-Theory and C^*-Algebras. Oxford University Press, New York (1993)
24. Zhang Lunchuan: Relation between Hereditary C^*-subalgebras and complemented submodules. Acta Math. Sin. (Chinese) **47**(4), 747–750 (2004)

Chapter 3
Quantum Markov Semigroups Based on Hilbert C^*-Modules

3.1 Module Operator Semigroups

In this chapter, we introduce the concept of module operator semigroup and give characterizations. Then the classical solution and mild solution of the abstract Cauchy problem are discussed in the module framework. Moreover, the focus of this chapter is to characterize Markov module operator semigroups and the corresponding operator-valued Dirichlet forms. The main results of this chapter include the Hille-Yosida type theorem, quantum Stone theorem, and Beurling-Deny criterion between a class of Markov module operator semigroups and operator-valued Dirichlet forms. At the end of this chapter, we prove the equivalence between hypercontractivity of a class of quantum Markov semigroups and associated logarithmic Sobolev inequality based on a probability gage space.

3.1.1 Background Knowledge

In statistical mechanics, it is important to characterize the stationarity of the correlation function of linear irreversible systems, whose main tool is based on the Stone theorem of semigroup theory on Hilbert space. The multidimensional linear irreversible system can form a Hilbert C^*-module. In order to characterize its correlation function matrix, we need to establish the corresponding module operator semigroup theory in the framework of Hilbert C^*-module.

Consider a class of quantum system $\{A_t\}_{t\in R^+}$, where $A_t = (\hat{a_1}(t), \hat{a_2}(t), \cdots, \hat{a_n}(t))$, in which $\hat{a_j}(t) = exp(-\frac{t\hat{H}}{ih})\hat{a_j}(0)exp(\frac{t\hat{H}}{ih}), t \in R^+$, i is the imaginary number unit, and each $\hat{a_j}(0)$ $(j = 1, 2, \cdots, n)$ is a bounded self-adjoint operator on some separable Hilbert space and represents an observable Hermitian quantity, and $\hat{H}$ stands for the nonperturbed Hamiltonian quantity. In what follows, let E be

L. Zhang, *Hilbert C*- Modules and Quantum Markov Semigroups*,
https://doi.org/10.1007/978-981-99-8668-2_3

the closed subspace in $B(H^n)$ generated linearly by $\{A_t\}_{t\in R^+}$. Define the $M_n(\mathbb{C})$-valued inner product on E by the following way:

$$< A_t; A_s >= \begin{pmatrix} < \hat{a}_1(t); \hat{a}_1(s) > & < \hat{a}_1(t); \hat{a}_2(s) >, & \cdots, & \hat{a}_1(t); \hat{a}_n(s) > \\ < \hat{a}_2(t); \hat{a}_1(s) > & < \hat{a}_2(t); \hat{a}_2(s) >, & \cdots, & < \hat{a}_2(t); \hat{a}_n(s) > \\ \cdots & \cdots & & \cdots \\ < \hat{a}_n(t); \hat{a}_1(s) > & < \hat{a}_n(t); \hat{a}_2(s) >, & \cdots, & < \hat{a}_n(t); \hat{a}_n(s) > \end{pmatrix},$$

where $< \hat{a}_i(t); \hat{a}_j(s) >$

$$= \beta^{-1}\int_0^\beta d\lambda Tr\{exp(-\beta\hat{H})exp(\lambda\hat{H})\hat{a}_i(t)exp(-\lambda\hat{H})\hat{a}_j(s)\}/Tr(exp(-\beta\hat{H}))$$

$(i,\ j = 1, 2, \cdots, n;\ t, s \in R^+)$, in which $\beta = \frac{1}{kT}$ is the reversible temperature, T is the absolute temperature, and k is the Boltzmann constant (see [35], formula 4.1.12). So E a is Hilbert $M_n(\mathbb{C})$-module. At Sect. 3.3, we shall characterize its correlation function and stationarity of the above quantum system $\{A_t\}_{t\in R^+}$.

3.1.2 Module Operator Semigroups and Related Concepts

Definition 3.1 Let E be a Hilbert A-module. A one-parameter family $T = \{T(t),\ 0 \le t < +\infty\}$, contained in $L(E)$ is called a module operator semigroup if

(1) $T(0) = I$, where I is the identity operator in $L(E)$;
(2) $T(t+s) = T(t) + T(s)\ (t,\ s \ge 0)$.

Moreover, module operator semigroup $T = \{T(t), 0 \le t < +\infty\}$ is called a contractive module semigroup if $||T(t)|| \le 1$ for all $t \ge 0$.

It is abbreviated as $T(t)$ or T_t without causing confusion.

In order to make a profound study for module operator semigroup, some form of continuity must be required:

Module operator semigroup $T(t)$ is called strongly continuous wherever

$$\lim_{t\to 0^+} T(t)x = x,$$

for all $x \in E$. Moreover, if $\{T(t)^*\}$ is also strongly continuous, then $T(t)$ is said to be strictly topologically continuous.

Definition 3.2 If $\{T_t,\ 0 \le t < +\infty\}$ is a module operator semigroup, then the A-linear operator G defined by

$$D(G) = \{x \in E : \lim_{t\to 0^+} \frac{T_t x - x}{t} exists\}$$

and

$$Gx = \lim_{t\to 0^+} \frac{T_t x - x}{t} = \frac{d^+ T_t x}{dt}|_{t=0} \ (x \in D(G))$$

is called the infinitesimal generator of the semigroup T_t, and $D(G)$ is the domain of G.

Remark 3.1 In Definition 3.1 and Definition 3.2, if the parameter set R^+ is replaced by the real number field $\mathbb{R}$ and if each T_t is unitary, then the concepts of unitary module operator group and its infinitesimal generator are obtained, respectively. It is easy to prove that for a unitary module operator group, its strong continuity is equivalent to its strict continuity.

According to the above definitions, strongly continuous module operator semigroup has the following basic properties:

Proposition 3.1 *Suppose that $\{T_t, 0 \le t < +\infty\} \subset L(E)$ is a strongly continuous module operator semigroup with infinitesimal generator G. Then the following conclusions hold:*

(1) There exist constants $\alpha \ge 0$ and $M \ge 0$ such that $||T_t|| \le Me^{\alpha t}$ $(t \in R^+)$;
(2) $\frac{d}{dt}T_t x = GT_t x = T_t Gx$ $(t \in R^+$ $x \in D(G))$;
(3) $\int_0^t T(s)xds \in D(G)$, and $G(\int_0^t T(s)xds) = T(t)x - x$ $(x \in E)$;
(4) $\int_s^t T(\tau)Gxd\tau = \int_s^t GT(\tau)d\tau = T(t)x - T(s)x$ $(x \in D(G))$;
(5) G is a densely defined closed operator;
(6) $\lim_{h\to 0} \frac{1}{h}\int_t^{t+h} T(s)xds = T(t)x$ $(x \in E)$.

Proof

(1) We can find a positive integer a such that $\{T(t), t \in [0, \ a]\}$ is bounded. If not, then there exists sequence $\{t_n\} \subset [0, +\infty)$ such that $\lim_{n\to+\infty} t_n = 0$ but $||T(t_n)|| \ge n$. Hence, from the uniform boundedness principle, there is a $x \in E$ such that $T(t_n)x$ is unbounded, which contradict to $\lim_{n\to+\infty} T(t_n)x = x$. Thus, there exist positive constants a and M such that

$$||T(t)|| \le M \ (t \in [0, a]).$$

Since $T(0) = 1$, it follows that $M \ge 1$. For any $t > 0$, $t = na + \varepsilon$, where $\varepsilon \in [0, a]$ and n is a nonnegative integer. Let $\alpha = a^{-1}\log M$, and then from the properties of semigroup,

$$||T(t)|| = ||T(\varepsilon)T(a)^n||$$

$$\le M^{n+1} \le M \cdot M^{\frac{t}{a}} = Me^{\alpha t}.$$

(2) If $x \in D(G)$, then

$$\lim_{h \to 0^+} \frac{T(h) - I}{h} T(t)x = \lim_{h \to 0^+} T(t) \left(\frac{T(h) - I}{h} \right) x = T(t)Gx.$$

Since $T(t)x \in D(G)$ and $GT(t)x = T(t)Gx$ from Definition 3.2, so that

$$\frac{d^+}{dt} T(t)x = GT(t)x = T(t)Gx.$$

Further, when $0 < h < t$, we have

$$\lim_{h \to 0^+} ||\frac{T(t)x - T(t-h)x}{h} - T(t)Gx||$$

$$\leq \lim_{h \to 0^+} ||T(t-h) \left(\frac{T(h)x - x}{h} - G(x) \right) || + \lim_{h \to 0^+} ||T(t-h)Gx - T(t)Gx||$$

$$\leq Me^{\alpha t} \lim_{h \to 0^+} ||\frac{T(h)x - x}{h} - Gx|| + \lim_{h \to 0^+} ||T(t-h)Gx - T(t)Gx|| = 0,$$

which shows that $\frac{d^-}{dt} T(t)x = T(t)Gx = GT(t)x$. Thus,

$$\frac{d}{dt} T(t)x = T(t)Gx = GT(t)x.$$

(3) If $x \in E$ and $h > 0$, then

$$\lim_{h \to 0^+} \frac{T(h) - I}{h} \int_0^t T(s)xds$$

$$= \lim_{h \to 0^+} \frac{1}{h} \int_0^t T(s+h)xds - \lim_{h \to 0^+} \frac{1}{h} \int_0^t T(s)xds$$

$$= \lim_{h \to 0^+} \frac{1}{h} \int_h^{t+h} T(s)xds - \lim_{h \to 0^+} \frac{1}{h} \int_0^t T(s)xds$$

$$= \lim_{h \to 0^+} \left(\frac{1}{h} \int_h^t T(s)xds + \frac{1}{h} \int_t^{t+h} T(s)xds - \frac{1}{h} \int_0^t T(s)xds \right)$$

$$= \lim_{h \to 0^+} \frac{1}{h} \int_t^{t+h} T(s)xds - \lim_{h \to 0^+} \frac{1}{h} \int_0^h T(s)xds$$

$$= T(t)x - T(0)x,$$

and therefore $\int_0^t T(s)xds \in D(G)$ and $G(\int_0^t T(s)xds) = T(t)x - x$.

(4) By integrating the formula (2) from s to t, we get

$$\int_s^t T(\tau)Gxd\tau = \int_s^t GT(\tau)xd\tau$$

$$= \int_s^t \int \frac{d}{d\tau}T(\tau)xd\tau = T(t)x - T(s)x.$$

(5) We first show that $\overline{D(G)} = E$. Since for any $x \in E$, $T(t)x$ is a strongly continuous function on $[0, +\infty)$, it follows that $T(t)x$ is Bochner integrable. Let $x_h = \frac{1}{h}\int_0^h T(s)xds$ if $h > 0$. Then using the conclusion (3) which we have proved above, we have $x_h \in D(G)$ and we have also

$$\lim_{h\to 0^+} x_h = \lim_{h\to 0^+} \frac{1}{h}\int_0^h T(s)xds$$

$$= T(0)x = x,$$

which shows that $\overline{D(G)} = E$.

We prove next that G is closed. Let $\{x_n\} \subset D(G)$ such that $x_n \to x_0$ and $Gx_n \to y_0$, $n \to \infty$. We must prove that $x_0 \in D(G)$ and $y_0 = Gx_0$. If it is proved, then G is a closed operator. Since from the conclusion (4), we have

$$T(h)x_n - x_n = \int_0^h T(s)Gx_nds \ (s \geq 0, h > 0).$$

Recall that each $T(s)$ is a bounded linear operator and $\{Gx_n\}$ is bounded; hence, $\{||T(s)Gx_n||\}$ is bounded, so that there exists a constant $M_s > 0$ (M_s depends on s but not on n) such that $||T(s)Gx_n|| \leq M$. Therefore, from the Lebesgue dominated convergence theorem that

$$\lim_{n\to\infty}(T(h)x_n - x_n) = \lim_{n\to\infty}\int_0^h T(s)Gx_nds$$

$$= \int_0^h T(s)\lim_{n\to\infty} Gx_nds = \int_0^h T(s)y_0ds.$$

Hence,

$$T(h)x_0 - x_0 = \int_0^h T(s)y_0ds.$$

Thus,

$$\lim_{h\to 0^+} \frac{T(h)x_0 - x_0}{h}$$

$$= \lim_{h\to 0^+} \frac{1}{h}\int_0^h T(s)y_0 ds$$

$$= T(0)y_0 = y_0,$$

so that $x_0 \in D(G)$ and $Gx_0 = y_0$.

(6) Since $T(t)x$ is a continuous function of t, it is Riemann integrable on $[t, t + h]$ $(h > 0)$ or $[t + h, t]$ $(h < 0)$. Since

$$||\frac{1}{h}\int_t^{t+h} T(s)xds - T(t)x||$$

$$= ||\frac{1}{h}\int_t^{t+h} (T(s)x - T(t)x)ds||$$

$$\leq \frac{1}{h}\int_t^{t+h} ||T(s)x - T(t)x||ds$$

$$\leq \sup_{t\leq s\leq t+h} ||T(s)x - T(t)x|| \to 0,\ h \to 0,$$

It follows that $\lim_{h\to 0} \frac{1}{h}\int_t^{t+h} T(s)xds = T(t)x$ $(x \in E)$. In particular, from the above, it implies that $\lim_{h\to 0} \frac{1}{h}\int_0^h T(s)xds = x$.

Remark 3.2 It can be proved that

$$\omega_0 = \inf_{t>0} \frac{\log ||T(t)||}{t} = \lim_{t\to\infty} \frac{\log ||T(t)||}{t}$$

exists, and if $\omega > \omega_0$, then there exists a constant M_ω such that

$$||T(t)|| \leq M_\omega e^{\omega t}\ (t \geq 0).$$

ω_0 is called the growth bound of strongly continuous module operator semigroup $T(t)$. In particular, the growth bound ω_0 of contractive module operator semigroups is 0.

3.1.3 Resolvents and Laplace Transforms

Let G be a (unbounded) module operator on Hilbert A-module E. Then a complex number λ is called a regular point of G if $(\lambda I - G)^{-1}$ is a bounded module operator. The set of all regular points of G is denoted by $\rho(G)$, which is called the resolvent set of G. In this case, $R(\lambda, G) = (\lambda I - G)^{-1}$ is said to be resolvent of G.

Suppose that T_t is a strongly continuous module operator semigroup. From Remark 3.2, if given any $\omega > \omega_0$, then there exists a constant $M_\omega > 0$ such that

$$||T(t)|| \leq M_\omega e^{\omega t} \ (t \geq 0),$$

where ω_0 is the growth bound of $\{T_t\}$. Therefore, for any fixed $s > 0$, when $t \in [0, \ s]$ and $Re\lambda > \omega > \omega_0$, we have

$$||e^{-\lambda t}T(t)x|| \leq ||e^{-\lambda t}T(t)|| \cdot ||x|| \leq M_\omega e^{-(Re\lambda - \omega)t} \cdot ||x||.$$

for all $x \ \in \ E$. Observe that $e^{-(Re\lambda-\omega)t}$ is Riemann integrable on $[0, \ \infty)$, so it is Bochner integrable. Thus, $e^{-\lambda t}T(t)x$ is Bochner integrable and $\lim_{s\to\infty}\int_0^s e^{-\lambda t}T(t)xdt$ exists. Moreover, let

$$\hat{T}(\lambda) = \int_0^\infty e^{-\lambda t}T(t)dt$$

$$= \lim_{s\to\infty}\int_0^s e^{-\lambda t}T(t)dt \in L(E).$$

Then $\hat{T}(\lambda)$ is called the Laplace transform of $T(t)$.

Theorem 3.1 *Let $\{T_t, \ 0 \leq t < +\infty\}$ be a strongly continuous module operator semigroup with infinitesimal generator G, and let ω_0 be its growth bound. If a complex number λ satisfies $Re\lambda > \omega_0$, then $\lambda \in \rho(G)$, and the following equation holds:*

$$R(\lambda, \ G)x = \hat{T}(\lambda)x = \int_0^\infty e^{-\lambda t}T(t)xdt \ (x \in E).$$

Proof Write $G_h = \frac{T(h)-I}{h}$. Since

$$G_h\hat{T}(\lambda)x = \int_0^\infty e^{-\lambda t}\frac{T(h)-I}{h}T(t)xdt$$

$$= \frac{1}{h}\int_0^\infty e^{-\lambda t}T(t+h)xdt - \frac{1}{h}\int_0^\infty e^{-\lambda t}T(t)xdt\iota$$

$$= \frac{e^{\lambda h}-1}{h}\int_0^\infty e^{-\lambda t}T(t)xdt - \frac{e^{\lambda h}}{h}\int_0^h e^{-\lambda t}T(t)xdt,$$

from which it implies that

$$\lim_{h\to 0} G_h \hat{T}(\lambda)\lambda x = \lambda \hat{T}(\lambda)x - x.$$

This shows that $\hat{T}(\lambda)x \in D(G)$ and $G\hat{T}(\lambda)x = \lambda\hat{T}(\lambda)x - x$. Therefore, $(\lambda I - G)\hat{T}(\lambda)x = x$. Thus $\hat{T}(\lambda)$ is the right inverse of $(\lambda I - G)$. Since

$$G_h \hat{T}(\lambda)x = \hat{T}(\lambda)G_h x,$$

it follows that

$$G\hat{T}(\lambda)x = \lim_{h\to 0^+} G_h \hat{T}(\lambda)x$$

$$\lim_{h\to 0^+} \hat{T}(\lambda)G_h x = \hat{T}(\lambda)Gx,$$

for all $x \in D(G)$. It results from the above proof that

$$\hat{T}(\lambda)Gx = \lambda\hat{T}(\lambda)x - x \ (x \in D(G)).$$

Hence, $\hat{T}(\lambda)(\lambda I - G)x = x$, which shows that $\hat{T}(\lambda)$ is also the left inverse of $(\lambda I - G)$. Thus, $\hat{T}(\lambda) = (\lambda I - G)^{-1} = R(\lambda, G)$.

Remark 3.3

(1) In Sect. 3.4, the resolvent $R(\lambda,\ G) = (\lambda I - G)^{-1}$ of G is written as G_λ for convenience;
(2) Similar to Theorem 3.1.7 in [3], $\hat{T}(\lambda) = R(\lambda, G)$ is an essential characterization of strongly continuous module operator semigroup, that is, if given a strongly continuous function $T : R^+ \to L(E)$, then the following conditions are equivalent:

   1° There is a module operator G and a constant ω_0 such that $(\omega_0, \infty) \subset \rho(G)$, and when $\lambda > \omega_0$, then $\hat{T}(\lambda)$ exists and $\hat{T}(\lambda) = R(\lambda,\ G)$;
   2° $T(t)$ is a strongly continuous module operator semigroup.

3.1.4 Hille-Yosida Type Theorem

In what follows, we shall consider the generation theory of strongly continuous contractive module operator semigroups: what kind of module operators can be infinitesimal generators of strongly continuous module operator semigroups? For this, similar to the classical case of C_0-semigroups on Banach spaces, we obtain the Hille-Yosida type theorem as follows.

Theorem 3.2 (Hille-Yosida type theorem) *Let E be a Hilbert A-module, and let G be a module operator with domain $D(G)$ in E. Then G becomes the infinitesimal generator of a strongly continuous contractive module operator semigroup if and only if G is a densely defined closed operator satisfying $(0, +\infty) \subset \rho(G)$ and the following inequality holds:*

$$||\lambda(\lambda I - G)^{-1}|| \leq 1 \ (\lambda > 0).$$

Proof Necessity. Suppose that G is the infinitesimal generator of a contractive module operator semigroup $T(t)$. From Proposition 3.1 (5), we see that G is a densely defined closed operator. Since T_t is contractive, its growth order $\omega_0 = 0$, it follows from Theorem 3.1 that $(0, +\infty) \subset \rho(G)$, and when $\lambda > 0$, we have

$$(\lambda I - G)^{-1}x = \int_0^\infty e^{-\lambda s}T(s)xds,$$

for all $x \in E$. Therefore,

$$\begin{aligned}||(\lambda I - G)^{-1}x|| &\leq \int_0^\infty e^{-\lambda s}||T(s)x||ds \\ &\leq \frac{1}{\lambda}||x||.\end{aligned}$$

Hence, $||\lambda(\lambda I - G)^{-1}x|| \leq ||x||$, from which it implies that $||\lambda(\lambda I - G)^{-1}|| \leq 1$.

Sufficiency. When $\lambda > 0$, let $G_\lambda = \lambda G(\lambda I - G)^{-1} = \lambda^2(\lambda I - G)^{-1} - \lambda I \in L(E)$. It is easy to check that $\{e^{tG_\lambda}, \ 0 < t < +\infty\}$ is a strongly continuous contractive module operator semigroup. Observe that

$$\begin{aligned}||e^{tG_\lambda}|| &= e^{-t\lambda}||exp[t\lambda^2(\lambda I - G)^{-1}]|| \\ &\leq e^{-t\lambda}exp[t\lambda^2||(\lambda I - G)^{-1}||] \leq 1.\end{aligned}$$

And since

$$||(\lambda I - G)^{-1}Gx|| \leq \frac{||Gx||}{\lambda} \to 0, \ \lambda \to \infty,$$

and

$$\lambda(\lambda I - G)^{-1}x - (\lambda I - G)^{-1}Gx = x,$$

for all $x \in D(G)$. It follows that $\lambda(\lambda I - G)^{-1}x \to x, \ \lambda \to \infty$. Recall that $\overline{D(G)} = E$ and the continuity of $(\lambda I - G)^{-1}$, so the above formula holds for all $x \in \overline{D(G)} = E$. Thus,

$$G_\lambda x = \lambda(\lambda I - G)^{-1}Gx \to Gx, \ \lambda \to \infty,$$

for all $x \in D(G)$. Since

$$e^{tG_\lambda}x - e^{tG_\mu}x = \int_0^1 \frac{d}{ds}\left(e^{tsG_\lambda}e^{t(1-s)G_\mu}x\right)ds$$

$$= \int_0^1 e^{tsG_\lambda}e^{t(1-s)G_\mu} \cdot t(G_\lambda x - G_\mu x)ds,$$

it implies that

$$||e^{tG_\lambda}x - e^{tG_\mu}x||$$

$$\leq \int_0^1 ||e^{tsG_\lambda}|| \cdot ||e^{t(1-s)G_\mu}|| \cdot ||t|| \cdot ||G_\lambda x - G_\mu x||ds$$

$$\leq t||G_\lambda x - G_\mu x|| \to 0,\ \lambda \to \infty, \mu \to \infty.$$

Set

$$T(t)x = \lim_{\lambda\to\infty} e^{tG_\lambda}x,\ (x \in E).$$

It is clear that $||T(t)|| \leq 1$ and $T(t)T(s) = T(t+s),\ T(0) = I$, where I is the identity module operator on E. Thus, $T(t)$ is a semigroup. Since

$$||e^{sG_\lambda}G_\lambda x|| \leq ||G_\lambda x|| \to ||Gx||, \lambda \to \infty\ ((x \in D(G))).$$

It results from Lebesgue dominated convergence theorem that

$$T(t)x - x = \lim_{\lambda\to\infty} e^{tG_\lambda}x - x$$

$$= \lim_{\lambda\to\infty} \int_0^t e^{sG_\lambda}G_\lambda x ds$$

$$= \int_0^t \lim_{\lambda\to\infty} e^{sG_\lambda}G_\lambda x ds$$

$$= \int_0^t T(s)Gx ds\ . \tag{3.1}$$

This shows that $\lim_{t\to 0^+} T(t)x = x(x \in D(G))$. By using the contractility of $T(t)$, the above formula holds for all x in $\overline{D(G)} = E$, so $\{T(t)\}_{t\in\mathbb{R}^+}$ is a strongly continuous module operator semigroup .

Further, if G' is the infinitesimal generator of $T(t)$, then from the above (3.1), we have $G' \supset G$. Hence, $D(G') \supset D(G)$ and $G'|_{D(G)} = G$. From the sufficiency hypothesis, $1 \in \rho(G)$, it follows by Theorem 3.1 that $1 \in \rho(G')$. Observe that $(I - G')D(G) = (I - G)D(G) = E$; hence, $D(G') = (I - G')^{-1}E = D(G)$, which shows that $G' = G$.

3.2 Abstract Cauchy Problem Based on Hilbert *A*-Modules

It is well known that the classical Cauchy problem is the initial value problem of differential equation. In this section, we consider the relationship between the solution of the abstract Cauchy problem based on Hilbert A-module and the strongly continuous module operator semigroup. Given a A-linear closed operator G on a Hilbert A-module E, consider the following linear homogeneous Cauchy problem: Let

$$(ACP_0) \quad \begin{cases} u'(t) = Gu(t)(t \geq 0), \\ u(0) = x. \end{cases}$$

where $x \in E$ and function $u: \ R^+ \to E$.

3.2.1 The Relationship Between Classical Solutions of Cauchy Problem and Strongly Continuous Module Operator Semigroups

Suppose that E is a Hilbert C^*-module over a C^*-algebra A. Write $C(R^+, E)$ for the set of all continuous functions from $\mathbb{R}^+$ to E, and $C^1(R^+, \ E)$ the set of all continuous differentiable functions from $\mathbb{R}^+$ to E.

Definition 3.3 A-linear function $u(t)$ in $C(R^+, \ E)$ is said to be a classical solution of the above Cauchy problem (ACP_0) if the following conditions hold:

(1) $u(t) \in D(G)$ $(t \geq 0)$;
(2) $u(t)$ is differentiable on $(0, \ \infty)$ and satisfies the equation (ACP_0).

In the following, we shall give the relationship theorem between the classical solution of the Cauchy problem (ACP_0) and strongly continuous module operator semigroups.

Theorem 3.3 *Let G be a densely defined module operator on a Hilbert A-module E. Then the following conditions are equivalent:*

(1) The operator G generates a strongly continuous module operator semigroup, that is, G is an infinitesimal generator of some strongly continuous module operator semigroup;

(2) *$\rho(G) \neq \varnothing$, and for any $x \in D(G)$, the above Cauchy problem (ACP_0) has a unique classical solution.*

Proof (1) $\Rightarrow$ (2) : Let G generate a strongly continuous module operator semigroup $\{T(t),\ 0 \leq t < +\infty\}$. If $x \in D(G)$, set $u(t) = T(t)x\ (t \in R^+)$. Then $u(t)$ is an E-valued continuous differentiable function, and therefore by Proposition 3.1 (2) that $u(\cdot)$ is a solution of Cauchy problem (ACP_0).

We next want to prove the uniqueness. If $v(\cdot)$ is a solution of the Cauchy problem (ACP_0), then for $0 \leq s \leq t < \infty$, we have

$$\frac{d}{ds}[T(t-s)v(s)] = T(t-s)Gv(s) - T(t-s)Gv(s) = 0,$$

which shows that $T(t-s)v(s)$ is independent of s, but is dependent only on t. Therefore, let $s = t$ and $s = 0$, respectively, in the above equation, and we can get

$$\begin{aligned} v(t) = T(0)v(t) &= T(t)v(0) \\ &= T(t)x = u(t). \end{aligned}$$

(2) $\Rightarrow$ (1) : Write $u_x(t)$ for the classical solution corresponding to x such that $u_x(0) = x$ for each fixed $x \in D(G)$. It is easy to see by the uniqueness condition of the solution that $u_x(t)$ is linear with respect to x. Since $\rho(G) \neq \varnothing$, it follows that G is a closed operator. Hence, $D(G)$ becomes a Banach space when endowed with the following graph norm:

$$||x||_G \triangleq ||x|| + ||Gx||,$$

for all $x \in D(G)$. For fixed $t \geq 0$, define a A-linear operator $T(t)$ on $D(G)$ given by

$$T(t)x = u_x(t)\ (x \in D(G)).$$

Now, to show $T(t)$ is bounded, for this, we shall construct a A-linear mapping $\Phi : D(G) \to C([0, t_0], D(G))$ for any fixed closed interval $[0,\ t_0]$ $(t_0 \geq t)$ such that $\Phi(x) = u_x(\cdot) \in C([0,\ t_0],\ D(G))$. We next prove that Φ is continuous. Since $D(G)$ is complete with graph norm, it suffices to prove that Φ has a closed graph by the closed graph theorem. If sequence $\{x_n\} \subset D(G)$ is convergent with graph norm on $D(G)$ and with uniform norm on $C([0,\ t_0],\ D(G))$, respectively, then

$$x_n \to x,\ \Phi(x_n) = u_{x_n} \to u, n \to +\infty.$$

Hence, $x \in D(G)$ since G is a closed operator, and therefore

$$\begin{aligned}\Phi(x_n) &= u_{x_n}(t) \\ &= x_n + \int_0^t Au_{x_n}(\tau)d\tau \\ &\to x + \int_0^t Gu(\tau)d\tau,\ n \to \infty.\end{aligned}$$

Since

$$\begin{aligned}\Phi(x)(t) &= u_x(t) \\ &= x + \int_0^t Gu(\tau)d\tau,\end{aligned}$$

and

$$\Phi(x_n) \to u, n \to \infty,$$

it follows that $\Phi(x)(t) = u(t)$, this shows that $\Phi(x) = u$. Thus, Φ is a closed operator. Combined with the above proof and the closed graph theorem, we get that Φ is bounded on $D(G)$, so that there exists a positive constant M such that

$$||\Phi(x)|| \leq M||x||_G\ (x \in D(G)).$$

Hence,

$$\max_{s\in[0,t_0]} ||u_x(s)||_G \leq M||x||_G.$$

Therefore,

$$\begin{aligned}||T(t)x||_G &= ||u_x(t)||_G \\ &\leq \max_{s\in[0,t_0]} ||u_x(s)||_G \leq M||x|||_G.\end{aligned}$$

Thus, through routine calculation, there exists a constant $M' > 0$ such that

$$||T(t)x|| \leq M'||x||\ (x \in D(G), t \in [0, t_0])\ . \tag{3.2}$$

Since $\overline{D(G)} = E$, it follows that the above formula holds for all $x \in E$, which shows that $T(t)$ is a bounded module operator. Moreover, let u be a classical solution corresponding to x. Then it is easy to check that $u(\cdot + s)(s \geq 0)$ is the classical solution corresponding to the initial value $u(s)$. From the uniqueness of the solution,

$$T(t+s)x = T(t)T(s)x \ (x \in D(G),\ t,\ s \geq 0),$$

for all $x \in E$ since $\overline{D(G)} = E$. Combined with the above formula (3.2), we see that $\{T(t)\}_{t \in R^+}$ is a strongly continuous module operator semigroup on E.

Finally, in order to prove G is the infinitesimal generator of $T(t)$, it suffices to show that if B is also the infinitesimal generator of $T(t)$, then $B = G$. From the sense of $T(t)$, if $x \in D(G)$, then $T(t)x = u_x(t)$, and therefore from the assumption (2), it follows that

$$\frac{d}{dt} T(t)x = GT(t)x \ (t \geq 0).$$

Hence, $(\frac{d}{dt}T(t))x = Gx$, which shows that $B \supset G$. Moreover, when $y \in D(G^2)$, it can be proved by the existence and uniqueness of the solution, $T(t)Gy = GT(t)y$. Let $Re\lambda > \omega_0$, where ω_0 is the growth bound of $T(t)$. From $B \supset G$ and the above equation, we have

$$e^{-\lambda t} GT(t)y = e^{-\lambda t} T(t)Gy$$

$$= e^{-\lambda t} T(t)By \ (y \in D(G^2)).$$

By integrating the above formula from 0 to ∞, we get that $GR(\lambda,\ B)y = R(\lambda, B)By$. Since $BR(\lambda,\ B)y = R(\lambda, B)By$, it follows that

$$GR(\lambda,\ B)y = BR(\lambda, B)y \ (y \in D(G^2)).$$

Observe that $BR(\lambda,\ B)$ is a bounded module operator, and since G is closed module operator with $\overline{D(G^2)} = E$, it follows that

$$GR(\lambda,\ B)y = BR(\lambda,\ B)y \ (y \in E).$$

Therefore, $D(G) \supset R(R(\lambda,\ B)) = D(B)$ and $G \supset B$. Hence, $G = B$.

3.2.2 *The Relationship Between Mild Solutions of Cauchy Problem and Strongly Continuous Module Operator Semigroups*

The requirements of classical solutions are too strong; in the following, we shall consider the mild solutions:

Definition 3.4 Let E be a Hilbert A-module. Then a continuous function $u : R^+ \to E$ is called a mild solution of the Cauchy problem (ACP_0) if it satisfies the following conditions:

$$\int_0^t u(s)ds \in D(G), \text{ and } G\int_0^t u(s)ds = u(t) - x, \ (t \geq 0).$$

Similar to the proof of Theorem 3.3, we can obtain the relation theorem between strongly continuous module operator semigroup and the mild solution of Cauchy problem as follows.

Theorem 3.4 *Let E be a Hilbert A-module, and let G be a A-linear closed module operator. Then the following conditions are equivalent:*

(1) The Cauchy problem (ACP_0) has a uniquely mild solution for each $x \in E$
(2) The operator G generates a strongly continuous module operator semigroup.

Remark 3.4 The necessary and sufficient condition for the mild solution $u : R^+ \to E$ of Cauchy problem (ACP_0) to be a classical solution is that u is a continuous differentiable function. Indeed, if the mild solution $u : R^+ \to E$ is a continuous differentiable function, let $t \geq 0$; for sufficiently small $h \neq 0$ and such that $t+h \geq 0$ (if $t = 0$ then let $h > 0$), then we have

$$\frac{1}{h}(u(t+h) - u(t)) = \frac{1}{h}G\int_t^{t+h} u(s)ds \in D(G).$$

Since G is a closed operator, it follows that

$$u(t) = \lim_{h\to 0}\frac{1}{h}\int_t^{t+h} u(s)ds \in D(G),$$

and $u'(t) = Gu(t)$.

We next consider the case where the parameter set is the entire set of real numbers. Suppose that E is a Hilbert A-module and G is a A-linear closed operator. Consider the following Cauchy problem:

$$(ACP_0(R)) \quad \begin{cases} u'(t) = Gu(t)(t \in R), \\ u(0) = x. \end{cases}$$

where $x \in E$. Function u in $C(R, \ E)$ is called the mild solution of Cauchy problem $ACP_0(R)$ if it satisfies that $\int_0^t u(s)ds \in D(G)$, and $G\int_0^t u(s)ds = u(t) - x \ (t \in \mathbb{R})$.

Theorem 3.5 *If module operator G generates a strongly continuous module operator semigroup T_+, at the same time, $-G$ generates a strongly continuous module operator semigroup T_-. Set*

$$(*)U(t) = \begin{cases} T_+(t), t \geq 0, \\ T_-(-t), t < 0. \end{cases}$$

Then $\{U(t), t \in R\} \subset L(E)$ is a strongly continuous module operator semigroup.

Proof Strong continuity and $U(0) = I$ are obvious. We next prove that $U(t+s) = U(t)U(s)$ $(t, s \in \mathbb{R})$. If $s,\ t \geq 0$, then

$$U(t+s) = T_+(t+s) = T_+(t)T_+(s) = U(t)U(s).$$

Similarly, if $s, t < 0$, then $U(t+s) = U(t)U(s)$ since $T_-(-t-s) = T_-(-t)T_-(-s)$. The only thing that need to show is that $U(t-s) = U(t)U(-s)$ $(t \geq 0,\ s \geq 0)$. Without loss of generality, let $0 \leq s \leq t$. For any $x \in E$, set

$$v(s) = T_+(t-s)x \ (s \in [0,\ t]).$$

Then for $0 \leq r \leq t$, we have

$$-G\int_0^r v(s)ds = -G\int_0^r\int_0^t T_+(t-s)xds$$

$$= -G\int_{t-r}^t T_+(s)xds$$

$$= T_+(t-r)x - T_+(t)x = v(r) - T_+(t)x.$$

Thus, v is a mild solution of the following problem:

$$(**)\begin{cases} v'(s) = -Gv(s),\ 0 < s \leq t, \\ v(0) = T_+(t)x. \end{cases}$$

From Theorem 3.1, when $Re\lambda > \omega(T_+)$, one can obtain that

$$R(\lambda, G) = \hat{T}_+(\lambda)$$

$$= \int_0^\infty e^{-\lambda t}T_+(t)xdt \ (x \in E).$$

When $Re\lambda > \omega(T_-)$. It implies that

$$R(\lambda, -G) = \hat{T}_-(\lambda)$$

$$= \int_0^\infty e^{-\lambda t}T_-(t)xdt,\ (x \in E).$$

Take $\lambda_0 = \max(\omega(T_+),\ \omega(T_-))$. We get

$$R(\lambda,\ G) = \int_0^\infty e^{-\lambda t}T_+(t)xdt,$$

and

$$R(\mu, -G) = \int_0^{\infty} e^{-\mu s} T_-(s)xds,$$

whenever $\lambda, \mu > \lambda_0$. The above proof shows that $R(\lambda, G)R(\mu, -G) = R(\mu, -G)R(\lambda, G)$. From the uniqueness of the Laplace transformation (see Proposition 3.1.5 in [3]), it is concluded that $T_+(t)T_-(s) = T_-(s)T_+(t)$ $(s, t \geq 0)$. On the other hand, it is easily checked that $T_-(s)T_+(t)x$ is also a mild solution to the Problem $(**)$. Hence, $v(s) = T_-(s)T_+(t)x$ from the uniqueness of mild solution. Therefore,

$$\begin{aligned} U(t-s)x &= T_+(t-s)x \\ &= v(s) = T_-(s)T_+(t)x \\ &= U(-s)U(t)x. \end{aligned}$$

3.2.3 *Characterizations of Strongly Continuous Module Operator Groups*

Definition 3.5 If E is a Hilbert A-module and G is a A-linear module operator, then G is called the infinitesimal generator of a strongly continuous module operator group if G and $-G$ generate strongly continuous module operator semigroups, respectively.

Theorem 3.6 *A-linear closed module operator G generates a strongly continuous module operator group, if and only if the Cauchy problem $ACP_0(R)$ has unique mild solution u for every $x \in E$.*

In this case, $u(t) = U(t)x$ $(t \in \mathbb{R})$, where $(U(t), t \in \mathbb{R})$ is the strongly continuous module operator group generated by G. In particular, if $x \in D(G)$, then $U(\cdot)x \in C^1(\mathbb{R}, E)$, $U(t)x \in D(G)$ $(t \in \mathbb{R})$ and $\frac{d}{dt}U(t)x = GU(t)x$ $(t \in \mathbb{R})$.

Proof If G generates a strongly continuous module operator group, then by simple computation, $U(\cdot)x$ is a mild solution of Cauchy problem $ACP_0(R)$. The rest can be obtained by properties of semigroup.

Conversely, if for every $x \in E$ the Cauchy problem $ACP_0(\mathbb{R})$ has a uniquely mild solution, then the following Cauchy problems have uniquely mild solutions, respectively:

$$(CP)_{\pm} \quad \begin{cases} u'(t) = \pm Gu(t), & t \geq 0, \\ u(0) = x, & . \end{cases}$$

Hence, G and $-G$ generate strongly continuous module operator semigroups from Theorem 3.4, respectively. Thus, G generates a strongly continuous module operator group from Theorem 3.5.

Definition 3.6 Let $T(t)$ be a strongly continuous module operator semigroup on a Hilbert A-module E. Continuous function $u : \mathbb{R} \to E$ is called a complete orbit of $T(t)$ if it satisfies that

$$u(t+s) = T(t)u(s) \ (t \geq 0, s \in \mathbb{R}).$$

Theorem 3.7 *Let $T(t)$ be a strongly continuous module operator semigroup with infinitesimal generator G. If $u : \mathbb{R} \to E$ is continuous and $u(0) = x$, then u is the mild solution of Cauchy problem $ACP_0(\mathbb{R})$ if and only if u is a complete orbit of $T(t)$.*

Proof Suppose that u is a complete orbit of $T(t)$. Let $x = u(0)$. Since for $t \geq 0$, $T(t)x = u(t)$, we have

$$G\int_0^t u(s)ds = G\int_0^t T(s)xds$$

$$= T(t)x - x = u(t) - x.$$

When $t < 0$. Since $T(-t)u(t) = u(0) = x$, and since

$$G\int_0^t u(r)dr = G\int_{-t}^0 u(r+t)dr$$

$$= -G\int_0^{-t} u(r+t)dr$$

$$= -G\int_0^{-t} T(r)u(t)dr$$

$$= u(t) - T(-t)u(t) = u(t) - x.$$

It follows that u is the mild solution of Cauchy problem $ACP_0(R)$.

Conversely, suppose that u is a mild solution of Cauchy problem $ACP_0(\mathbb{R})$ and $x = u(0)$. Then for $t \geq 0,\ s \in \mathbb{R}$, we have

$$G\int_0^t u(r+s)dr = G\int_s^{s+t} u(r)dr = u(t+s) - u(s),$$

which implies that $u(\cdot + s)$ is the mild solution of Cauchy problem ACP_0 corresponding to the initial value $x = u(s)$, so that $u(t + s) = T(t)u(s)$ $(t \geq 0)$.

Observe that if a strongly continuous module operator semigroup $T(t)$ can be extended to a group of strongly continuous module operators, then every $T(t)$ is invertible. In fact, the inverse proposition is true.

Theorem 3.8 *Let G be the infinitesimal generator of a strongly continuous module operators semigroup $T(t)$. If there exists $t_0 > 0$ such that $T(t_0)$ is invertible and then G can generate a strongly continuous module operator group.*

Proof We first prove that if $t > 0$. then $T(t)$ is invertible. Assume that $T(t)x = 0$. Choose n such that $nt_0 > t$. Then we have

$$T(nt_0)x = T(nt_0 - t)T(t)x = 0.$$

Since $T(nt_0) = T(t_0)^n$ is invertible, it follows from what we have proved that $x = 0$, and therefore $T(t)$ is injective. Let $x = T(nt_0 - t)T(nt_0)^{-1}y$ for each fixed $y \in E$. Then we have

$$T(t)x = T(t)T(nt_0 - t)T(nt_0)^{-1}y = y,$$

which shows that $T(t)$ is also surjective. Further, define a function U from $\mathbb{R}$ to $L(E)$ as follows: when $t \geq 0$, $U(t) = T(t)$, and when $t < 0$, $U(t) = T(-t)^{-1}$. Then it is easy to check that $U(t + s) = U(t)U(s)$ $(s, t \in \mathbb{R})$, that is, $U(t)$ is a semigroup. Given $t_0 \in \mathbb{R}$, let $t_1 > \max\{-t_0,\ 0\}$. Then for every $x \in E$, we have

$$\lim_{t \to t_0} U(t)x = T(t_1)^{-1} \lim_{t \to t_0} T(t + t_1)x = U(t_0)x,$$

which shows that $U(t)$ is strongly continuous.

Finally, in order to show that $U(t)$ is a group, it suffices to prove that $-G$ is the infinitesimal generator of $U(-t)$. Since for any $x \in D(G)$, we have

$$\lim_{t \downarrow 0}(U(-t)x - x)$$

$$= \lim_{t \downarrow 0} U(-1)(\frac{1}{t}T(1 - t)x - T(1)x)$$

$$= T(1)^{-1}(-GT(1)x) = -Gx.$$

On the other hand, if $x \in E$ such that $y = \lim_{t\downarrow 0}(U(-t)x - x)$ exists, then

$$
\begin{aligned}
-y &= \lim_{t\downarrow 0} \frac{1}{t}(x - U(-t)x) \\
&= \lim_{t\downarrow 0} \frac{1}{t}(x - T(t)^{-1}x) \\
&= \lim_{t\downarrow 0} \frac{1}{t}T(t)(x - T(t)^{-1}x) \\
&= \lim_{t\downarrow 0} \frac{1}{t}(T(t)x - x),
\end{aligned}
$$

from which it implies that $x \in D(G)$ and $-Gx = y$, so that $-G$ is the infinitesimal generator of $U(-t)$.

3.3 Quantum Stone Theorem and Its Application

3.3.1 Quantum Stone Theorem

Stone theorem is the representation theorem of unitary operator group with one parameter. It has important applications in quantum mechanics, statistical mechanics, and group representation theory, as mentioned at the beginning of this chapter, In statistical mechanics, multidimensional irreversible systems can form a Hilbert C^*-module whose inner product belongs to a matrix algebra. In order to characterize its correlation function and the stationarity of the process, the classical Stone theorem needs to be extended to the case of Hilbert C^*-module.

Definition 3.7 Closed densely defined module operator T on a Hilbert A-module E is called regular if T has densely defined adjoint T^* and the range $R(I + T^*T)$ of $I + T^*T$ is also densely defined, where I is the identity on E.

Proposition 3.2 (See Lemma 9.8 in [24]) *If T is a densely defined self-adjoint module operator, then T is regular if and only if $R(T+iI) = E$ and $R(T-iI) = E$, where i is the imaginary unit and I is the identity on E.*

Theorem 3.9 (Quantum Stone theorem) *Let E be a Hilbert A-module, and let G be a A-linear module operator. Then G becomes the infinitesimal generator of some strictly continuous unitary module operator group $\{U_t\}_{t\in R} \subset L(E)$ if and only if iG is a regular self-adjoint module operator. In this case, let $T = -iG$; formally, it can be written as $U_t = e^{itT}$ $(t \in \mathbb{R})$.*

In order to prove Theorem 3.9, we first give the concept of dissipative module operator and a Lumer-Phillips type theorem as lemma.

Definition 3.8 Module operator T on a Hilbert A-module E is called a dissipative module operator if it satisfies the following condition:

$$||(\lambda I - T)x|| \geq \lambda||x|| \ (x \in D(T), \lambda > 0).$$

Lemma 3.1 (Lumer-Phillips type theorem) *Let T be a densely defined module operator on a Hilbert A-module E. Then T is the infinitesimal generator of a strongly continuous contractive module operator semigroup $(T_t, t \in \mathbb{R}^+)$ if and only if T is a dissipative module operator and there exists a positive number λ_0 such that $R(\lambda_0 I - T) = E$.*

Proof Necessity. Let T be the infinitesimal generator of strongly continuous contractive module operator semigroup $\{T_t\}_{t \in R^+}$. From Theorem 3.2, we see that $\rho(T) \supset (0, +\infty)$ and $R(\lambda I - T) = E \ (\lambda > 0)$. Since

$$||x|| = ||(\lambda I - T)^{-1}(\lambda I - T)x||$$
$$\leq \frac{1}{\lambda}||(\lambda I - T)x|| \ (x \in D(T)).$$

It follows that $||(\lambda I - T)x|| \geq \lambda||x|| \ (x \in D(T))$, namely, T is a dissipative module operator.

Sufficiency. Since T is a dissipative module operator, $||(\lambda I - T)x|| \geq \lambda||x|| \ (x \in D(T), \lambda > 0)$. Let $\lambda = \lambda_0$ in the above inequality; it follows that $||(\lambda_0 I - T)x|| \geq \lambda_0||x|| \ (x \in D(T))$, and therefore from the assumption $R(\lambda_0 I - T) = E$ that $(\lambda_0 I - T)^{-1}$ is bounded. Hence, $\lambda_0 \in \rho(T)$, which shows that $\rho(T) \neq \varnothing$ and therefore T is closed.

We next show that T generates a strongly continuous contractive module operator semigroup; it suffices to prove that $R(\lambda I - T) = E \ (\lambda > 0)$. If this is proved, then $(\lambda I - T)^{-1}$ is bounded and $||\lambda(\lambda I - T)^{-1}|| \leq 1$ since T is dissipative. Therefore, the conclusion can be drawn from Theorem 3.2. For this, consider the set $\Gamma = \{\lambda : 0 < \lambda < \infty \text{ and } R(\lambda I - T) = E\}$. Since $\lambda_0 \in \Gamma$, $\Gamma \neq \varnothing$. Given any $\lambda \in \Gamma$, then $||\lambda x - Tx|| \geq \lambda||x|| \ (x \in D(T))$ since T is dissipative, from which it implies that $\lambda \in \rho(T)$. Recall that $\rho(T)$ is open; hence, there exists a neighborhood $O_\lambda (\subset \rho(T)$ of λ such that $O_\lambda \cap \mathbb{R} \subset \Gamma$, which shows that Γ is an open subset in $(0, +\infty)$.

Furthermore, assume that a sequence $\lambda_n \subset \Gamma$ and $\lambda_n \to \lambda > 0$ as $n \to \infty$. Since $R(\lambda_n I - T) = E, n = 1, 2, \cdots$, it follows that for every $y \in E$, there exists $x_n \in D(T)$ corresponding to λ_n such that $\lambda_n x_n - Tx_n = y$, from which it implies that $||x_n|| \leq \lambda_n^{-1}||y|| \leq C$ since T is dissipative, where the constant C is independent of n. Hence,

$$\lambda_m||x_n - x_m|| \leq ||\lambda_m(x_n - x_m) - T(x_n - x_m)||$$

$$= ||\lambda_m x_n - \lambda_m x_m - Tx_n + Tx_m||$$

$$= ||\lambda_m x_n - Tx_n - (\lambda_m x_m - Tx_m)||$$

$$= ||\lambda_m x_n - \lambda_n x_n + (\lambda_n x_n - Tx_n) - (\lambda_m x_m - Tx_m)||$$

$$= ||\lambda_m x_n - \lambda_n x_n + y - y||$$

$$= ||(\lambda_m - \lambda_n)x_n|| \leq C|\lambda_n - \lambda_m|.$$

The above proof shows that $\{x_n\}$ is a Cauchy sequence in E. We write x for the limit of $\{x_n\}$, when $n \to \infty$. It results from the above equation that $\lambda_n x_n - Tx_n = y$, so $Tx_n \to \lambda x - y$ as $n \to \infty$. Since T is a closed operator, then $x \in D(T)$ and $\lambda x - Tx = y$, and therefore $R(\lambda I - T) = E$. Thus, $\lambda \in \Gamma$, that is, Γ is also a closed subset in $(0, +\infty)$. Hence, $\Gamma = (0, +\infty)$.

The proof of Theorem 3.9 If G is the infinitesimal generator of U_t, then

$$Gx = \lim_{t\to 0^+} \frac{(U_t - I)x}{t}$$

$$= \lim_{t\to 0^+} U_t \cdot \frac{(I - U_{-t})x}{t}$$

$$= - \lim_{t\to 0^+} U_t \cdot \frac{(U_t^* - I)x}{t} = -G^*x,$$

for all $x \in D(G)$, where G^* is the adjoint of G, so that $G = -G^*$, and therefore iG is a self-adjoint module operator. From the proof of Lemma 3.1, we have

$$R(\lambda I - G) = E, \text{ and } R(\lambda I - G^*) = E \ (\lambda > 0).$$

Let $\lambda = 1$ in the above equations; then one can obtain $R(I - G) = E$ and $R(I + G) = R(I - G^*) = E$. It follows that

$$R((iG) + iI) = R(i(I + G)) = R(I + G) = E,$$

and

$$R((iG) - iI) = R(i(G - I)) = R(G - I) = R(I - G) = E.$$

Thus, iG is a regular module operator by Proposition 3.2.

Conversely, if iG is a self-adjoint regular module operator, then $G = -G^*$. Since

$$< (\lambda I - G)x,\ (\lambda I - G)x >$$

$$= \lambda^2 < x,\ x > -\lambda < x,\ Gx > -\lambda < Gx,\ x > + < Gx,\ Gx >$$

$$= \lambda^2 < x,\ x > + < Gx,\ Gx >$$

$$\geq \lambda^2 < x,\ x >,$$

for all $x \in D(G)$ and $\lambda > 0$. It follows that

$$||(\lambda I - G)||^2$$

$$= || < (\lambda I - G)x,\ (\lambda I - G)x > ||$$

$$\geq \lambda^2 || < x,\ x > ||$$

$$= \lambda^2 ||x||^2,$$

for all $x \in D(G)$ and $\lambda > 0$. Hence, $||(\lambda I - G)|| \geq \lambda ||x||$, which shows that G is dissipative. Because iG is a self-adjoint regular module operator, then G is densely defined and closed with $R(iG - iI) = E$ from Proposition 3.2. Therefore,

$$R(I - G) = R(i(iG - iI)) = R(iG - iI) = E.$$

Thus, the claim is proved by Lemma 3.1.

Theorem 3.9 becomes the classical Stone theorem when Hilbert C^*-module reduces to Hilbert space.

Proposition 3.3 *A linear operator G on a Hilbert space H becomes the infinitesimal generator of a strongly continuous unitary operator group if and only if iG is a self-adjoint operator.*

Proof From [22] we see that the self-adjointness of iG is equivalent to $R(iG \pm iI) = H$, from which it shows that G is regular, since H is itself a Hilbert C^*-module over the complex field $\mathbb{C}$. Thus, the conclusion can be obtained by Theorem 3.9.

Example 3.1 Given a von Neumann algebra A and a σ-weak continuous one parameter unitary operator group $\{U_t = e^{itT}, t \in \mathbb{R}\} \subset A$, then the following spectral decomposition holds:

$$U_t = \int_{-\infty}^{+\infty} e^{it\lambda} dE_\lambda,$$

where $\{E_\lambda : -\infty < \lambda < +\infty\} \subset L(A) = A$ is the resolution of the identity for T.

In fact, A is itself a W^*-module with A-valued inner product $< a, b >= a^*b$ $(a,\ b \in A)$. By using Theorem 3.9 and combining with spectral decomposition of unbounded self-adjoint operator, the decomposition formula is obtained.

3.3.2 *Spectral Decomposition of a Class of Stationary Quantum Processes*

Definition 3.9 Let E be a Hilbert A-module. A family $\{\xi_t\}_{t\in\mathbb{R}} \subset E$ is said to be a quantum stochastic process. Moreover, the function defined by the following way:

$$C(\cdot,\ \cdot) : \mathbb{R} \times \mathbb{R} \to A,\ C(s,\ t) =< \xi_s,\ \xi_t >\in A$$

is called the correlation function corresponding to the above quantum stochastic process $\{\xi_t\}_{t\in\mathbb{R}}$.

Remark 3.5 Similar to the classical case, the above correlation function $C(\cdot,\ \cdot)$ is a positive definite operator-valued function. Indeed, for any $s_1,\ s_2,\ \cdots,\ s_n \in \mathbb{R}$ and any $a_1,\ a_2,\ \cdots,\ a_n \in A$, we have

$$\sum_{i,\ j=1}^{n} a_i^* C(s_i,\ s_j) a_j$$

$$= \sum_{i,\ j=1}^{n} a_i^* < \xi_{s_i},\ \xi_{s_j} > a_j$$

$$=< \sum_{i=1}^{n} \xi_{s_i} a_i,\ \sum_{i=1}^{n} \xi_{s_i} a_i >\geq 0.$$

Definition 3.10 The meaning of $\{\xi_t\}_{t\in\mathbb{R}}$ is the same as in Definition 3.9. Then $\{\xi_t\}_{t\in\mathbb{R}}$ is called stationary if there exists a one-parameter unitary group $\{U_t\}_{t\in\mathbb{R}} \subset L(E_0)$ such that $\xi_{t+s} = U_t\xi_s$ $(t,\ s \in \mathbb{R})$, where E_0 is the closed submodule of E generated by $\{\xi_t\}_{t\in\mathbb{R}}$. In this case, $C(s,\ s+t) =< \xi_s,\ \xi_{t+s} >=< \xi_0, \xi_t >= C(0,\ t)$ $(t,\ s \in \mathbb{R})$.

The following properties characterize the relationship between stationary quantum processes and their correlation functions.

Proposition 3.4 *$\{\xi_t\}_{t\in\mathbb{R}}$ is stationary if and only if $C(t,\ s)$ depends only on $s - t$.*

Proof The necessary is trivial from Definition 3.10. Sufficiency is proved as follows: Suppose that $C(t, s)$ depends only on $s - t$. We have $C(t + p, s + p) = C(t, s)$ for all $t, s, p \in \mathbb{R}$. It follows that

$$||\sum_{i=1}^{n} \xi_{p+t_i}||^2 = || < \sum_{i=1}^{n} \xi_{p+t_i}, \sum_{i=1}^{n} \xi_{p+t_i} > ||$$

$$= || \sum_{i,\ j=1}^{n} < \xi_{p+t_i}, \xi_{p+t_j} > || = || \sum_{i,\ j=1}^{n} C(p + t_i, p + t_j)||$$

$$= || \sum_{i,\ j=1}^{n} C(t_i, t_j)|| = || < \sum_{i=1}^{n} \xi_{t_i}, \sum_{i=1}^{n} \xi_{t_i} > ||$$

$$= ||\sum_{i=1}^{n} \xi_{t_i}||^2,$$

for all $\xi_{t_i} \in \{\xi_t\}_{t\in\mathbb{R}}$, $i = 1, 2, \cdots, n$. Hence, there exists an isometry U_p from a dense subspace of E_0 to E_0 given by $\xi_t \to \xi_{p+t}$ $(t \in \mathbb{R})$ for each fixed $p \in \mathbb{R}$, where E_0 is the closed submodule of E generated by $\{\xi_t\}_{t\in\mathbb{R}}$. Thus, U_p can be extended to an isometry from E_0 to E_0, which is still denoted by U_p for convenience, and it is easy to check that $U_p^*\xi_t = \xi_{t-p}$. Therefore, from Theorem 1.2, we see that $U_p \in L(E_0)$ is unitary for each fixed $p \in \mathbb{R}$, from which one can obtain a family of unitary module operators $\{U_p\}_{p\in\mathbb{R}} \subset L(E_0)$. From the construction of U_p, it is easily verified that $U_0 = I$, $U_t U_s = U_{t+s}$ and $\xi_t = U_t\xi_0$, which indicates that $\{U_p\}_{p\in\mathbb{R}}$ is the group of unitary module operators with one parameter that satisfies the claim.

Remark 3.6 It is easy to check that the following conditions are equivalent:

(1) Stationary quantum stochastic process $\{\xi_t\}_{t\in\mathbb{R}}$ is mean-square continuous with respect to the norm on the Hilbert A-module E;
(2) $\{U_t\}_{t\in\mathbb{R}}$ is strictly continuous
(3) $\{U_t\}_{t\in\mathbb{R}}$ is strongly continuous, that is, $\lim_{t\to 0} ||U_t\xi_s - \xi_s|| = 0$ for every $\xi_s \in \{\xi_t\}_{t\in\mathbb{R}}$.

It is well known that in classical stochastic process, if given positive definite kernel $C(s, t)$ $(s, t \in T$, T is the index set), then there exists a Gaussian system $\{\xi_t\}_{t\in T}$ that has zero mean value and its second moment (i.e., correlation function) $E(\overline{\xi_s}\xi_t) = C(s, t)$, vice versa.

In the framework of Hilbert A-modules, the following results will be obtained.

Theorem 3.10 *Let $C(s, t) : \mathbb{R} \times \mathbb{R} \to A$ be a A-valued function. Then $C(s, t)$ is positively defined if and only if there exist a Hilbert A-module E and a quantum process $\{\xi_t\}_{t\in\mathbb{R}} \subset E$ such that $C(s, t) =< \xi_s, \xi_t >$.*

Proof The sufficiency has been shown in Definition 3.10. Now we just need to prove the necessity. Suppose that $C(s, t)$ is positively defined. Define A-valued function given by

$$a_t(\cdot) : \mathbb{R} \to A, \ a_t(s) = C(s, t)a,$$

for each fixed $a \in A$ and each fixed $t \in \mathbb{R}$. Write E_0' for the linear space generated linearly by $\{a_t : a \in A, \ t \in \mathbb{R}\}$. Then E_0' is a pre-Hilbert A-module when endowed with the following A-valued inner product:

$$< f, g >= \sum_{i, j=1}^{n} \lambda_i^* \mu_j (a^i)^* C(t_i, t_j) b^j,$$

for all $f = \sum_{i=1}^{n} \lambda_i a_{t_i}^i$ and $g = \sum_{j=1}^{n} \mu_j b_{t_j}^j \in E_0'$, where a^i and b^j in A $(i, j = 1, 2, \cdots, n)$, λ_i and μ_j in $\mathbb{C}$ $(i, j = 1, 2, \cdots, n)$. The completion of E_0' is a Hilbert A-module, which is denoted by E_0.

Now we can construct a quantum process that satisfies our requirements in the following way: Let $\xi_t : A \to E_0$ such that $\xi_t(a) = a_t$, and let $\xi_t^* : E_0 \to A$ such that $\xi_t^*(a_s) = a_s(t) = C(t, s)a$ for all t in $\mathbb{R}$ and a in A. Since

$$< \xi_t(b), \ a_s >=< b_t, \ a_s >$$

$$= b^* C(t, s)a =< b, \ \xi_t^* a_s >,$$

for all $a, b \in A$. It follows that $\xi_t \in L(A, E_0)$ $(t \in \mathbb{R})$. Set $E = L(A, E_0)$. Then E is a Hilbert A-module with the A-valued inner product given by $< f, g >= f^* g$ $(f, g \in E)$. Hence, $\{\xi_t\}_{t \in \mathbb{R}} \subset E$ is just the quantum process that we wanted. Indeed, from the above proof, we have

$$< \xi_t, \ \xi_s > (a) = \xi_t^* \xi_s(a)$$

$$= \xi_t^*(a_s) = a_s(t)$$

$$= C(t, s)a,$$

for all $a \in A$, so that $< \xi_t, \ \xi_s >= C(t, s)$.

Using Theorem 3.9 and Theorem 3.10, the spectral decomposition of the correlation function for a class of stationary quantum stochastic processes can be described as follows.

Theorem 3.11 *Let E be a Hilbert A-module and $\{\xi_t\}_{t\in\mathbb{R}} \subset E$ a mean-square continuous stationary quantum process with correlation function $C(t)$. Then $C(t)$ has spectral decomposition as follows:*

$$C(t) = C(s,\ t+s) = \int_{-\infty}^{+\infty} e^{i\lambda t} d < \xi_0,\ E_\lambda \xi_0 >,$$

where the resolution of the identity $\{E_\lambda : -\infty < \lambda < +\infty\}$ is in $L(\widetilde{E})$, in which $\widetilde{E}$ is the selfdual extension of E.

Proof From Definition 3.10 and the assumption, there exists a one-parameter strictly continuous unitary module operator group $\{U_t\}_{t\in\mathbb{R}} \subset L(E)$ such that $\xi_t = U_t \xi_0$ for all $t \in \mathbb{R}$, from which it implies that $C(s,\ s+t) =< \xi_0,\ U_t \xi_0 >$. We next work in $L(\widetilde{E})$. By Theorem 3.9, there exists a resolution of the identity $\{E_\lambda : -\infty < \lambda < +\infty\} \subset L(\widetilde{E})$ such that $U_t = \int_{-\infty}^{+\infty} e^{it\lambda} dE_\lambda$. Therefore,

$$C(s,\ t+s) =< \xi_0,\ U_t \xi_0 >= \int_{-\infty}^{+\infty} e^{it\lambda} d < \xi_0,\ E_\lambda \xi_0 > .$$

Example 3.2 We now consider the quantum system $\{A_t\}_{t\in\mathbb{R}}$ given at the beginning of this chapter. It is easy to check that

(1) $< \hat{a}_i(t); \hat{a}_j(t+s) >=< \hat{a}_i(0),\ \hat{a}_j(s) >$;
(2) $< \hat{a}_i(t); \hat{a}_j(s) >=< \hat{a}_j(s),\ \hat{a}_i(t) >\ \ (t,\ s \in \mathbb{R}),\ i,\ j = 1,\ 2,\ \cdots,\ n.$

It implies that $< A_t;\ A_{t+s} >=< A_0;\ A_s >$. Thus, $\{A_t\}_{t\in\mathbb{R}}$ is a stationary quantum process. From Theorem 3.11, its correlation function $C(t)$ has the following spectral decomposition:

$$C(t) = \int_{-\infty}^{+\infty} e^{it\lambda} d < \xi_0,\ E_\lambda \xi_0 >,$$

where the resolution of the identity $\{E_\lambda : -\infty < \lambda < +\infty\}$ is in $L(\widetilde{E})$, in which $\widetilde{E}$ is the selfdual extension of E.

3.4 A Class of Markov Module Operator Semigroups and Operator-Valued Dirichlet Forms

The theory of commutative Dirichlet forms, originated with the work of Beurling and Deny [4], had been developed especially by Fukushima [17] and Silverstein [33]. The study of Dirichlet forms and the associated Markov semigroups in the noncommutative setting of a C^*-algebra with a semifinite trace τ on it was pioneered by S. Albeverio and R. Hoegh- Krohn [2]. Since then, the theory of noncommutative

Dirichlet forms and Markov semigroups based on various noncommutative probability spaces has flourished. It has been recognized that noncommutative Dirichlet form theory shares a flavor of geometry in the sense of Connes' noncommutative geometry [13].

The various noncommutative Dirichlet forms that are mentioned above are all scalar-valued; it is natural to expect an extension of the theory of noncommutative Dirichlet forms to the operator-valued cases. Using this idea and method to describe corresponding n-positive (completely positive) Markov semigroups would be more natural. Observe that Hilbert W^*- bimodules are the generalizations of Hilbert spaces and von Neumann algebras, respectively, and they can be regarded as a bridge between Hilbert spaces and von Neumann algebras also. Therefore, it is appropriate to develop operator-valued Dirichlet forms and associated Markov module operator semigroups on Hilbert W^*-bimodules.

In this section, we shall characterize a class of weak $*$-continuous Markov module operator semigroups, noncommutative symmetric operator-valued Dirichlet forms, and the Beurling- Deny criterion between them, based on a class of Hilbert W^*-bimodule E which is generated by a II_1 factor A, by means of trace preserving faithful normal conditional expectation $\Phi : A \to B$, where B is a subfactor of A. For a further deeper discussion based on a broader framework, refer to the author's recent paper [36].

3.4.1 Characterization of Operator-Valued Quadratic Forms

The main theme of this subsection is to characterize the one-to-one correspondence between closed densely defined nonnegative operator-valued quadratic forms and closed densely defined self-adjoint nonnegative module operators, based on the Hilbert W^*-bimodule E which is generated by a II_1 factor A.

Suppose that A is a II_1 factor and B is its subfactor. Then there exists a unique trace preserving faithful normal conditional expectation $\Phi : A \to B$, so that A becomes a pre-Hilbert C^*-module over the von Neumann algebra B with the B-valued inner product $< x, y >= \Phi(x^*y)$ for all $x, y \in A$. Therefore, the closure of A with respect to the weak $*$ topology is a Hilbert W^*-module over the von Neumann algebra B (see the discussion preceding Remark 1.12, and see the Appendix), which is denoted by E. Since the canonical map $j : a \to a,\ A \to E$ is injective, we can identity a and $j(a)$ in E for all $a \in A$, and the unit 1 of A is the unit of E also. The operation $* : a \to a,\ A \to A$ has weak $*$ continuous extension onto E since A is dense in E.

Furthermore, since $b\ x \in A$ and $< b\ x,\ y >=< x,\ b^*y >$ for all $x,\ y \in A,\ b \in B$. It is easy to check that $b\ e \in E$ and $< b\ e, f >=< e, b^*f >$ for all $e, f \in E,\ b \in B$. Thus, E becomes a Hilbert W^*-bimodule with involution. In addition, a submodule D of E is a $*$ invariant submodule in case $x^* \in D$ whenever $x \in D$.

For example, let A be a hyperfinite II_1 factor. By the uniqueness of the hyperfinite II_1 factor (in the sense of isometric isomorphism), we can regard it as the completion of $\cup_n M_{2^n}(\mathbb{C})$ under the normalized tracial state τ. The notation τ_n means the normalized tracial state on $\cup_n M_{2^n}(\mathbb{C})$. The mapping

$$a \to \begin{bmatrix} a & 0 \\ \\ 0 & a \end{bmatrix}$$

is the embedding of $\cup_n M_{2^n}(\mathbb{C})$ as a subalgebra of $\cup_n M_{2^{n+1}}(\mathbb{C})$.

For each n, we have a conditional expectation map $\Phi_n : A \to M_{2^n}(\mathbb{C})$, where Φ_n is defined by

$$\Phi_n(a_1 \otimes a_2 \otimes \cdots \otimes a_{n+k}) = \left(\prod_{j=1}^{k} \tau(a_{n+j}) \right) a_1 \otimes a_2 \cdots \otimes a_n.$$

Since $\cup_{j\geq 0} M_{2^j}(\mathbb{C})$ is dense in A, we then extend Φ_n to A. Hence, we get an operator-valued noncommutative probability space $(A,\ M_{2^n}(\mathbb{C}),\ \Phi_n)$. It follows that a Hilbert W^*-bimodule E over $M_{2^n}(\mathbb{C})$ is constructed.

We now introduce the concept of B-valued symmetric quadratic form on the above Hilbert W^*-bimodule E over B:

Definition 3.11 B-valued sesquilinear form $\varepsilon(,\ )$ on the Hilbert W^*-bimodule E is a mapping $\varepsilon : D(\varepsilon) \times D(\varepsilon) \to B$ with the following properties:

(1) $\varepsilon(x,\ ya + zb) = \varepsilon(x,\ y)a + \varepsilon(x,\ z)b$;
(2) $\varepsilon(xa + yb,\ z) = a^*\varepsilon(x,\ z) + b^*\varepsilon(y,\ z)$;
(3) $\varepsilon(ax,\ y) = \varepsilon(x,\ a^*y)$,

where $D(\varepsilon)$ is a $*$ invariant submodule of E.

$\varepsilon(,)$ is called a symmetric sesquilinear form provided that $\varepsilon(x,\ y)^* = \varepsilon(y,\ x)$ for all $x,\ y \in D(\varepsilon)$. Correspondingly, $\varepsilon[x] = \varepsilon(x,\ x)$ is said to be a B-valued symmetric quadratic form. Furthermore, $\varepsilon[x]$ is called a B-valued nonnegative definite quadratic form whenever $\varepsilon[x] \geq 0$ for all $x \in D(\varepsilon)$. In addition, $\varepsilon(,\ )$ is said to be τ-real sesquilinear if it satisfies that $\overline{\tau\varepsilon(x,\ y)} = \tau\varepsilon(x^*,\ y^*)$ for all $x,\ y \in E$, where $-$ is denoted as the complex conjugate and τ is the unique faithful finite normal tracial state on B.

Remark 3.7

(1) In particular, the B-valued inner product $<,\ >$ on E is a B-valued positive definite sesquilinear form;
(2) Similar to the proof of Hilbert space case, it is easy to check that if the above sesquilinear $\varepsilon(,\ )$ is nonnegative definite, then it is symmetric and the Cauchy-Schwarz type inequality holds:

$$\varepsilon(y,\ x)\varepsilon(x,\ y) \leq ||\varepsilon(y,\ y)||\varepsilon(x,\ x).$$

In view of Definition 3.11, B-valued nonnegative definite quadratic form $\varepsilon[x] = \varepsilon(x, x)$ determines a B-valued inner product on $D(\varepsilon)$, which induces a norm on $D(\varepsilon)$: $||x||_1 = ||\varepsilon_1[x]||^{1/2}$, where $\varepsilon_1(x, y) = \varepsilon(x, y) + < x, y >$ for all $x, y \in D(\varepsilon)$. It is clear that $|| \cdot ||_1$ is stronger than $|| \cdot ||$. If the unit ball of $D(\varepsilon)$ is complete in the weak $*$ topology induced by the family of seminorms $\{p_{\varphi,\xi} = |\varphi(\varepsilon_1(\cdot, \xi))| : \varphi \in B_*, \xi \in D(\varepsilon)\}$, then $D(\varepsilon)$ is a W^*-module itself, and in this case, $\varepsilon[,]$ is called a closed form. Furthermore, if $D(\varepsilon)$ is dense in E with respect to the weak $*$-topology on E, then $\varepsilon[,]$ is called a densely definite closed form.

Proposition 3.5 *There is a one-to-one correspondence between the family of closed densely defined nonnegative definite B-valued quadratic forms and the family of closed densely defined self-adjoint nonnegative module operators on E.*

Proof Suppose that $\varepsilon(,)$ on $D(\varepsilon)$ is a closed densely defined nonnegative definite B-valued sesquilinear form. Set $\tilde{D} = \{x \in D(\varepsilon)$: there exists $\lambda_x > 0$ such that $||\varepsilon_1(x, y)|| \leq \lambda_x ||y||$, for all $y \in D(\varepsilon)\}$. Since $D(\varepsilon)$ is dense in E, the bounded module map $y \to \varepsilon_1(x, y)$, $D(\varepsilon) \to B$, can be extended uniquely to a bounded module map on E for each fixed $x \in \tilde{D}$. Since E is selfdual, there exists a unique $x' \in E$ such that $\varepsilon_1(x, y) =< x', y >$ for all $y \in D(\varepsilon)$. Therefore, we can define a module operator $\tilde{L}$ from $\tilde{D}$ to E such that $\tilde{L}x = x'$. Thus, $\varepsilon_1(x, y) =< \tilde{L}x, y >$ for all $x \in \tilde{D}$ and $y \in D(\varepsilon)$. Hence, $\tilde{L}$ is a closed positive modular operator, since $\varepsilon_1(,)$ is a closed positive form.

We next want to prove that $\tilde{L}$ is weak $*$ densely defined. For this, it suffices to show that if $\varepsilon_1(x, y) = 0$ for all $x \in \tilde{D}$, then $y = 0$. Since $\varepsilon_1(x, y) =< \tilde{L}x, y >$, it is enough to prove that $R(\tilde{L}) = E$, where $R(\tilde{L})$ is the range of $\tilde{L}$. Indeed, since for any $y \in E$, we have

$$|| < y, x > || \leq ||y|| \cdot ||x|| \leq ||y|| \cdot ||x||_1, \ x \in D(\varepsilon).$$

Therefore, the map $x \to < y, x >$ is a bounded module map from $D(\varepsilon)$ to B. Since $D(\varepsilon)$ is selfdual itself as a Hilbert W^*-module with respect to the new inner product $\varepsilon_1(,)$, there exists a unique $x'' \in D(\varepsilon)$ such that $< y, x >= \varepsilon_1(x'', x)$, $x \in D(\varepsilon)$. By the construction of $\tilde{L}$, it follows that $x'' \in \tilde{D}$ and $\tilde{L}x'' = y$. Hence, $R(\tilde{L}) = E$. Therefore, $y = 0$ from $\varepsilon(x, y) =< \tilde{L}x, y >= 0$ for all $x \in \tilde{D}$.

In the end, to show that $\tilde{L}$ is self-adjoint, for this, it suffices to prove that $\tilde{L}$ is symmetric and satisfies that $D(\tilde{L}^*) \subset D(\tilde{L})$. On the one hand, since $\varepsilon(,)$ is symmetric, that is, $\varepsilon(x, y) = \varepsilon(y, x)^*$, it follows that

$$< \tilde{L}x, y >= \varepsilon(x, y) = \varepsilon(y, x)^* =< \tilde{L}y, x >^* =< x, \tilde{L}y > .$$

Hence, $\tilde{L}$ is symmetric. On the other hand, by the meaning of $\tilde{L}^*$, for a given $x \in D(\tilde{L}^*)$, there exists a $x' \in E$ such that $< x', y >=< x, \tilde{L}y >$ for all $y \in D(\tilde{L})$.

By the conclusion proved in the preceding part, $R(\tilde{L}) = E$. Thus, there exists a $x'' \in D(\tilde{L})$ such that $x' = \tilde{L}x''$. Therefore,

$$< x'', \tilde{L}y >=< \tilde{L}x'', y >=< x', y >=< x, \tilde{L}y > .$$

This shows that $x = x'' \in D(\tilde{L})$. Put $L = \tilde{L} - I$, where I is the identity module map on E. Hence, L is a closed self-adjoint nonnegative definite module operator on E, and we have

$$\varepsilon(x, y) = \varepsilon_1(x, y)- < x, y >=< \tilde{L}x, y > - < x, y >=$$

$$< (\tilde{L} - I)x, y >=< Lx, y >, x \in D(\tilde{L}) = D(L), y \in D(\varepsilon).$$

Conversely, given any closed densely defined self-adjoint nonnegative module operator L, let $\varepsilon(x, y) =< Lx, y >$. Then it is easy to see that $\varepsilon[x] =< Lx, y >$ is the closed densely defined nonnegative B-valued quadratic form which is associated with L.

Remark 3.8 It is well known that closed submodules of Hilbert C^*-modules need not be orthogonally complemented, and this is the cause of all the difficulties that follows. In the Hilbert space case, a densely defined closed operator has a densely defined adjoint operator, but we would not expect that to hold for module operators on Hilbert C^*-modules, since even bounded module operators need not be adjointable. So the property of having a densely defined adjoint has to be built for further analysis. For the further development of Hilbert C^*-module theory, the concept of regular module operator is introduced in [24] (see Definition 3.7). Regular module operators maintain almost all the important properties of closed densely defined operators on Hilbert spaces, among them, from [24] Proposition 10.6 a closed densely defined symmetric module operator L on E is regular if and only if the submodules $R(L \pm iI)$ are complemented in E. In particular, if the above L is self-adjoint, from [24] Lemma 9.8, L is regular if and only if the module operators $L \pm iI$ are surjective. It follows that the Cayley transform of L can be constructed as follows:

$$C_L x = (L - iI)(L + iI)^{-1}x, x \in D(L).$$

The corresponding Cayley inverse transform is that

$$Lx = i(I + C_L)(I - C_L)^{-1}x, \ x \in D(L).$$

Therefore, L is self-adjoint if and only if C_L is unitary.

Next, we shall consider the spectral calculus of self-adjoint regular module operators on the above Hilbert W^*-bimodule E over subfactor B. Recall that a family $\{E_\lambda : -\infty < \lambda < +\infty\}$ of projection module operators on E is said to be spectral family if $E_\lambda E_\mu = E_\lambda$, $\lambda \le \mu$, $\lim_{\lambda' \downarrow \lambda} E_{\lambda'}x = E_\lambda x$, $\lim_{\lambda \to -\infty} x =$

$0, \lim_{\lambda\to+\infty} x = x,\ x \in E$. Given a self-adjoint regular module operator L on E, its Cayley transform C_L is a unitary module operator in $L(E)$. Note that $L(E)$ is a von Neumann algebra; using the spectrum decomposition representation of C_L in $L(E)$, there exists a spectral family $\{E_\lambda : -\infty < \lambda < +\infty\} \subset L(E)$ such that $C_L x = \int_{-\infty}^{+\infty} \frac{\lambda-i}{\lambda+i} dE_\lambda x,\ x \in E$, in the sense of weak $*$-convergence of approximating Riemann sums, that is, $f(< C_L x, y >) = \int_{-\infty}^{+\infty} \frac{\lambda-i}{\lambda+i} df(< E_\lambda x,\ y >$ for all $y \in E$ and for all $f \in B_*$. Then combining the Cayley inverse transform, similar to the discussion of Hilbert space case, one can obtain the spectrum decomposition representation of $L : Lx = \int_{-\infty}^{+\infty} \lambda dE_\lambda x, x \in D(L)$ in the sense of weak $*$-convergence of approximating Riemann sums. Furthermore, if the above L is nonnegative, then $\sqrt{L} = \int_{-\infty}^{+\infty} \sqrt{\lambda} dE_\lambda$.

In particular, if the above II_1 subfactor B is reduced to the matrix algebra $M_n(\mathbb{C})$, then every closed submodule of E is complemented [25]; it follows from [24] Proposition 10.6 that each closed densely defined self-adjoint module operator is regular. Therefore, by the spectrum decomposition of unbounded self-adjoint regular modular operators as mentioned above, combining Proposition 3.5, we see that closed densely defined nonnegative definite matrix-valued quadratic form $\varepsilon(,)$ can be represented as $\varepsilon(x,\ y) =< \sqrt{L}x,\ \sqrt{L}y >$ for all $x,\ y \in D(\varepsilon)$. Thus, $D(\varepsilon) = D(\sqrt{L})$.

3.4.2 A Class of Markov Module Operator Semigroups

In order to develop Markov module operator semigroups and corresponding operator-valued Dirichlet forms, we first briefly review the Markov semigroups and the scalar-valued Dirichlet forms based on probability gage space:

Let $(A,\ \tau)$ be a probability gage space; thus, A is a finite von Neumann algebra and τ is a faithful, normal trace on it. For $1 \le p < \infty$, $L^p(A,\ \tau)$ is the completion of A with respect to the norm $||x||_p = (\tau(|x|^p))^{\frac{1}{p}},\ x \in A$, and $L^\infty(A, \tau) = A$ equipped with the operator norm. These spaces share all the functional analytic features of the classical L^p-spaces, such as the uniform convexity for $p \in (0,\ \infty)$, duality between $L^p(A, \tau)$ and $L^{p'}(A,\ \tau)$ with $p^{-1} + p'^{-1} = 1$, and Riesz-Thorin interpolation, Hӧlder's and Clarkson's inequalities.

Markov semigroup and its associated Dirichlet form can be developed based on the standard form $(A,\ L^2(A,\ \tau), L^2_+(A,\ \tau),\ J)$ of the von Neumann algebra A, where $L^2_+(A,\ \tau)$ is a closed convex cone in $L^2(A, \tau)$, inducing an anti-linear isometry J (the modular conjugation) on $L^2(A, \tau)$ which is the extension of the involution $a \to a^*$ of A. The subspace of J-invariant elements (called real) will be denoted by $L^2_h(A,\ \tau)$.

When a is real, the symbol $a \wedge 1$ will denote the Hilbert projection onto the closed and convex subset $\{a \in L^2_+(A,\ \tau) : a \le 1\}$, where 1 is the unit of A.

Definition 3.12 Let $T = \{T(t), 0 \leq t < +\infty\}$ be a strongly continuous semigroup of bounded linear operators defined on $L^\infty(A, \tau)$. Then

(1) it is symmetric, if $\tau(T_t(x)y) = \tau(xT_t(y))$;
(2) it is Markovian, if $0 \leq x \leq 1$ implies that $0 \leq T_t(x) \leq 1$;
(3) it is conservative, if $T_t(1) = 1$ for all $t \geq 0$;
(4) it is completely Markovian, if $\{T_t \otimes I_n\}$ is Markovian on $L^\infty(A, \tau) \otimes M_n(\mathbb{C})$ for all $n \in \mathbb{N}$, where I_n is the identity map on $M_n(\mathbb{C})$.

Remark 3.9 From [14] Proposition 3.1 and [20] Theorem 3.3, the (completely) Markov semigroup on $L^\infty(A, \tau)$ can be extended to (completely) Markov semigroup on $L^p(A, \tau)$.

Definition 3.13 A closed densely defined and nonnegative quadratic form $(\varepsilon, D(\varepsilon))$ on $L^2(A, \tau)$ is said to be

(1) real if for $a \in D(\varepsilon)$ then $J(a) \in D(\varepsilon)$ and $\varepsilon[J(a)] = \varepsilon[a]$;
(2) a Dirichlet form if it is real and $\varepsilon[a \wedge 1] \leq \varepsilon[a]$, for $a \in D(\varepsilon) \cap L^2_h(A, \tau)$. Furthermore, it is conservative in case $1 \in D(\varepsilon)$;
(3) a regular Dirichlet form if in addition $A \cap D(\varepsilon)$ is norm dense in A and is also dense in $D(\varepsilon)$ with respect to the graph norm: $|||x|||_1^2 = \varepsilon[x] + \tau(|x|^2)$;
(4) a completely Dirichlet form if the canonical extension $(\varepsilon^n, D(\varepsilon^n))$ to $L^2(A, \tau) \otimes M_n(\mathbb{C}), \tau^n)$: $\varepsilon^n[[a_{ij}]_{i,j=1}^n] := \sum_{i,j=1}^n \varepsilon[a_{ij}]$, is a Dirichlet form for all $n \geq 1$, where $[a_{ij}]_{i,j=1}^n \in D(\varepsilon^n) := D(\varepsilon) \otimes M_n(\mathbb{C})$, and $\tau^n = \tau \otimes tr_n$ is the faithful, normal trace on the von Neumann algebra $A \otimes M_n(\mathbb{C})$, where tr_n is a normalized trace on $M_n(\mathbb{C})$.

Proposition 3.6 (see [2] Lemma 2.3, [12] Proposition 4.5, Proposition 4.10, [14] and Proposition 2.12) *Let $(\varepsilon, D(\varepsilon))$ be a closed densely defined nonnegative real quadratic form. Then the following statements are equivalent:*

(1) $(\varepsilon, D(\varepsilon))$ is a Dirichlet form;
(2) For every real-valued Lipschitz function $\varphi : \mathbb{R} \to \mathbb{R}$, which satifies $|\varphi(t) - \varphi(s)| \leq c_\varphi |t - s|$ for all $t, s \in \mathbb{R}$ and $\varphi(0) = 0$, where c_φ is a positive constant, we have $\varepsilon[\varphi(x)] \leq c_\varphi^2 \varepsilon[x]$ whenever $x \in D(\varepsilon) \cap L^2_h(A, \tau)$.
Furthermore, if $(\varepsilon, D(\varepsilon))$ is conservative, then the above items (1) *and* (2) *are equivalent to the following item:*
(3) $\varepsilon(1, x) \geq 0$ for all $x \in D(\varepsilon) \cap L^2_+(A, \tau)$, and $\varepsilon[|x|] \leq \varepsilon[x]$ for all $x \in D(\varepsilon) \cap L^2_h(A, \tau)$.

The following conclusion is the well-known Beurling-Deny type criterion in the noncommutative context.

Theorem 3.12 (Beurling-Deny) *(see [2] Theorem 2.7, 2.8; [12] Theorem 4.11 and [14] Theorem 3.3) Let $T_t = e^{-tL}$ be a strongly continuous symmetric semigroup with infinitesimal generator L, and $\varepsilon[x] =< \sqrt{L}x, \sqrt{L}x >$ for $x \in D(\varepsilon) = D(\sqrt{L})$ be the associated quadratic form. Then the following are equivalent:*

(1) The form ε is a (completely) Dirichlet form;
(2) The semigroup $T_t = e^{-tL}$ is (completely) Markovian.

Remark 3.10 Suppose that $\varepsilon[\ ,\ ]$ is a Dirichlet form. By [14] Proposition 1.2, the following inequality holds true:

$$\varepsilon[|x|] \leq 2\ \varepsilon[x]\ for\ all\ x \in D(\varepsilon).$$

When $x \in D(\varepsilon) \cap L^2(A,\ \tau)$, from Proposition 3.6 item (2), it is easy to check that the coefficient 2 on the right-hand side of the above inequality can be replaced by 1, that is, $\varepsilon[|x|] \leq \varepsilon[x]$.

Now we are going to introduce Markov module operator semigroups based on the above W^*-module E over II_1-factor B.

Definition 3.14 Module operator semigroup $T = \{T(t),\ 0 \leq t < +\infty\}$ on the above W^*-module E is called weak $*$ continuous if it satisfies the following statements:

(1). Map $t \to T(t)x,\ [0, +\infty) \to E$, is weak $*$ continuous for all $x \in E$, i.e., $t \to \phi(< T(t)x, y >)$ is continuous for all $\phi \in B_*$ and $y \in E$;
(2). $x \to T(t)x,\ E \to E$, is weak $*$–weak $*$ continuous for all $t \geq 0$.

Suppose that $T = \{T(t),\ 0 \leq t < +\infty\}$ on E is a weak $*$ continuous module operator semigroup. Then the linear module operator L defined by

$$D(L) = \{x \in E :\ \lim_{t\to 0} \frac{T(t)x - x}{t} exists\}$$

with respect to the above weak $*$ topology and

$$Lx = \lim_{t\to +0} \frac{T(t)x - x}{t} = \frac{d^+T(t)x}{dt}|_{t=+0}\ for\ all\ x \in D(L)$$

is the infinitesimal generator of the semigroup $T = \{T(t),\ 0 \leq t < +\infty\}$, and in this case, $D(L)$ is the domain of L.

Remark 3.11

(1) From Proposition 3.1 and Definition 3.14, we see that the above infinitesimal generator L is weak $*$ dense defined and weak $*$ closed.
(2) $T = \{T(t),\ 0 \leq t < +\infty\}$ is said to be regular provided that L is regular.

In Hilbert space theory, there exists a one-to-one correspondence between the family of strong continuous symmetric contraction semigroups and the family of closed densely defined self-adjoint nonpositive operators.

The above conclusion can be generalized to the weak $*$ continuous symmetric contraction and regular module operator semigroup cases:

Proposition 3.7 *Suppose that E is the above introduced Hilbert W^*-bimodule E over subfactor B. Then there exists a one-to-one correspondence between the family $\mathfrak{T}$ of weak $*$ continuous symmetric contraction regular module operator semigroups and the family $\mathfrak{L}$ of closed densely defined self-adjoint nonpositive regular module operators on E.*

Proof Assume that $T_t = e^{tL}, t \geq 0$ is a weak $*$ continuous symmetric contraction regular module operator semigroup with infinitesimal generator L. Then L is a weak $*$ closed and weak $*$ densely defined, self-adjoint, and nonpositive regular module operator on E. Set a mapping

$$\Phi : \mathfrak{T} \to \mathfrak{L}, e^{tL} \to L.$$

The above map Φ is well defined and is injective, since $\{T_t\}_{t\geq 0} = \{e^{tL}\}_{t\geq 0}$ and L are mutually determined. The next key is to prove that it is surjective. Let L be a weak $*$ closed and weak $*$ densely defined, self-adjoint, and nonpositive definite regular module operator on E. In order to prove that L is an infinitesimal generator of a weak $*$ continuous symmetric contraction regular module operator semigroup, using Theorem 3.2, we only need to prove the inequality $||\alpha(\alpha I - L)^{-1}|| \leq 1$. Indeed, from Remark 3.8, there exists a spectrum family $\{E_\lambda : 0 \leq \lambda \leq +\infty\} \subset L(E)$, in the sense of weak $*$-convergence of approximating Riemann sums. It follows that

$$-Lx = \int_{[0,\ +\infty)} \lambda dE_\lambda x,\ x \in D(L).$$

Therefore,

$$f(<\alpha(\alpha I - L)^{-1}x,\ \alpha(\alpha I - L)^{-1}x>) =$$

$$\int_{[0,\ +\infty)} \frac{\alpha}{(\alpha+\lambda)^2} df(<E_\lambda x,\ x>) \leq ||x||^2,\ \forall f \in B_*,\ x \in E.$$

This yields that $||\alpha(\alpha I - L)^{-1}|| \leq 1$.

Remark 3.12 Suppose that $T_t = e^{tL}, t \geq 0$ is a weak $*$ continuous symmetric contraction regular module operator semigroup with infinitesimal generator L and the associated resolvents $\{G_\alpha,\ \alpha > 0\}$. Define approximation symmetric forms $\varepsilon^{(t)}$ and $\varepsilon^{(\beta)} (\beta > 0)$ as follows:

(1) $\varepsilon^{(t)}(x,\ y) = \frac{1}{t} < x - T_t x,\ y >,\ x,\ y \in E;$
(2) $\varepsilon^{(\beta)}(x,\ y) = \beta < x - \beta G_\beta x,\ y >,\ x,\ y \in E.$

Using the spectral family just as in Remark 3.8, similar to the proof in [2] Lemma 1.3.4, then we have

$$\varepsilon[x] = \lim_{t\to 0} \varepsilon^{(t)}[x], \quad \varepsilon[x] = \lim_{\beta\to\infty} \varepsilon^{(\beta)}[x],$$

in the sense of weak $*$ convergence, respectively.

To proceed further discussion, we shall introduce the positive cone and order structure on the above E such that it becomes an order Hilbert W^*-bimodule.

Define the self-adjoint part E_h and the positive cone E^+ of E are the closure of A_h and A^+ in E with respect to the weak $*$ topology respectively, where A_h and A^+ are the self-adjoint part and the positive cone of A, respectively. It is easy to check that self-adjoint part E_h and the positive cone E^+ of E have the following basic properties:

(1) For any $x \in E_h$, there exist uniquely $x^+,\ x^- \in E^+$ such that $x = x^+ - x^-$ and $< x^+,\ x^- >= 0$;
(2) $\tau(< x,\ y >) \geq 0$ for all $x,\ y \in E^+$, and $E^+ = \{x \in E^+ : \tau(< x,\ y >) \geq 0$ for all $y \in E^+\}$, where τ is the faithful finite normal trace on B.

Define $x \geq y$, if $x - y \in E^+$ for $x,\ y \in E_h$. Hence, it introduces an order structure on E_h such that E becomes an order Hilbert W^*-bimodule.

Now we can introduce the concept of a class of Markov module operator semigroup as below:

Definition 3.15 A weak $*$ continuous, symmetric, and contraction module operator semigroup $\{T_t = e^{tL}\}_{t\geq 0} \subset L(E)$ is called a Markov module operator semigroup, if for any $x \in E_h$ and $0 \leq x \leq 1$, then $0 \leq T_t x \leq 1$, where 1 is the unit of E. In this case, every $T_t (t \geq 0)$ is called a Markov module operator.

From the meaning of Definition 3.15 and combined with Theorem 3.1 we can get directly the following conclusion:

Proposition 3.8 *The notations are the same as in Remark 3.12. Then the following conditions are equivalent:*

(1) $\{T_t\}_{t\geq 0}$ is a Markov module operator semigroup;
(2) each αG_α Markov module operator for all $\alpha > 0$.

Example 3.3 Suppose that A is a II_1 factor, and $\{P_t\}_{t\geq 0}$ is a τ-symmetric Markov semigroup on A, that is, it is made up of positive normal linear maps on A such that $P_t 1 \leq 1$ and the map $t \to P_t(a),\ [0,\ +\infty) \to A$, is continuous with respect to the σ-weak topology on A for each $a \in A$, where 1 is the unit of A. $\{P_t\}_{t\geq 0}$ is said to be τ-symmetric, if $\tau(bP_t(a)) = \tau(P_t(b)a)$ for all $a,\ b \in A$ and all $t \geq 0$. Furthermore, it is said to be n-positive, if its canonical extension $P_t \otimes I_n$ to the C^*-algebra $M_n(A) = A \otimes M_n(\mathbb{C})$:

$$(P_t \otimes I_n)\left([a_{ij}]_{i,\ j=1}^n\right) = [P_t(a_{ij})]_{i,\ j=1}^n,$$

is Markovian, where I_n is the unit of matrix algebra $M_n(\mathbb{C})$. If for any natural number n the Markov semigroup $\{P_t\}_{t\geq 0}$ is n-positive, then it is said to be completely Markov semigroup. Similarly define the τ_n symmetry.

In the following, we shall consider a τ_n-symmetric n-positive quantum Markov semigroup $\{P_t\}_{t\geq 0}$. Set

$$T_t^{(n)} = P_t \otimes I_n : A \otimes M_n(\mathbb{C}) \to A \otimes M_n(\mathbb{C}).$$

Then $\{T_t^{(n)}\}_{t\geq 0}$ is a quantum Markov semigroup on $A \otimes M_n(\mathbb{C})$. Note that $A \otimes M_n(\mathbb{C})$ is a pre-Hilbert W^*-bimodule over the matrix algebra $M_n(\mathbb{C})$ with the matrix-valued inner product $< x, y >= \Phi(x^*y)$, where $\Phi : A \otimes M_n(\mathbb{C}) \to M_n(\mathbb{C})$, $\Phi(a \otimes M) = \tau(a)M$ for all $a \in A$ and all $M \in M_n(\mathbb{C})$, is a trace preserving faithful normal conditional expectation. The closure of $A \otimes M_n(\mathbb{C})$ with respect to the weak $*$-topology is a W^*- bimodule, which is denoted by E_n.

From the above observation, it is easy to check that the Markov module operator semigroup $\{T_t^{(n)}\}_{t\geq 0}$ can be extended by weak $*$ continuity to a Markov regular module operator semigroup on Hilbert W^*-bimodule E_n; when no confusion can arise, we shall use the same symbol to denote the extension. Furthermore, if the above $\{P_t\}$ is completely positive, then $T_t^{(2^n)} = P_t \otimes I_{2^n}$ is a Markov module operator semigroup on $A \otimes M_{2^n}(\mathbb{C})$ for all natural number n.

Let E be the weak $*$ closure of pre-Hilbert W^*-bimodule $A \otimes_{alg} B$ with B-valued inner product $< a \otimes b, c \otimes d >= \tau(a^*c)b^*d, a \otimes b, c \otimes d \in A \otimes_{alg} B$. Hence, E is a Hilbert W^*-bimodule over the hyperfinite II_1 factor B. It follows that $\{T_t\}_{t\geq 0}$ can be extended by weak $*$ continuity to a Markov module operator semigroup on the Hilbert W^*-bimodule E.

3.4.3 *Operator-Valued Dirichlet Forms*

This subsection is devoted to characterize the Beurling-Deny correspondence between a class of Markov module operator semigroups and the associated operator-valued Dirichlet forms based on the above Hilbert W^*-bimodule E over a II_1 factor B.

Definition 3.16 When $x,\ y \in E_h$, define $x \vee y = y + (x - y)^+;\ x \wedge y = y - (x - y)^-$.

It is easy to prove that the above relations have the following basic properties:

(1) $x \vee y = y \vee x,\ x \wedge y = y \wedge x$;
(2) $x + y = x \vee y + y \wedge x$;
(3) $|x - y| = x \vee y - x \wedge y$.

Remark 3.13 In the commutative function space $L^2(X, \mu)$, where X is locally compact space with σ-finite measure μ on it. Then $x \vee y = \sup(x, y)$; $x \wedge y = \inf(x, y), \forall x, y \in L_h^2(X, \mu)$.

Now we can introduce the operator-valued Dirichlet form on E.

Definition 3.17 Let $(\varepsilon[\ ],\ D(\varepsilon))$ be a weak $*$ closed and weak $*$ densely defined, nonnegative defined B-valued τ-real quadratic form on the above W^*-bimodule E. Then it is called a B-valued Dirichlet form, if it satisfies the following statement:

$$x \in D(\varepsilon) \cap E_h \Longrightarrow x \wedge 1 \in D(\varepsilon), \text{ and } \tau(\varepsilon[x \wedge 1]) \leq \tau(\varepsilon[x]).$$

The above B-valued Dirichlet form possesses the following properties:

Proposition 3.9

(1) If $x \in D(\varepsilon)$, then $x^+ \in D(\varepsilon)$ and $\tau(\varepsilon[x^+]) \leq \tau(\varepsilon[x])$;
(2) If $x \in D(\varepsilon)$, then $|x| \in D(\varepsilon)$ and $\tau(\varepsilon[|x|]) \leq \tau(\varepsilon[x])$.

Proof

(1) Since the trace τ on B is faithful, E is an inner product space with respect to the inner product defined by

$$(x,\ y) = \tau(< x,\ y >) \ for\ all\ a, y \in E.$$

The completion of E is a Hilbert space with respect to the norm induced by the above inner product, which is denoted by H_τ. Since the map $i : E \to H_\tau, x \to x$, is a linear order-preserving injection, so that H_τ is an order Hilbert space, it follows that the selfdual-positive cone H_τ^+ and the real subspace H_τ^h in H_τ are the closure of E^+ and E_h, respectively. Hence, the B-valued Dirichlet form $\varepsilon(,)$ on E can be extended to a scalar-valued Dirichlet form $\varepsilon_\tau(,)$ on H_τ such that $\varepsilon_\tau(x, y) = \tau(\varepsilon(x, y))$ for all $x,\ y \in E$.

For each $x \in D(\varepsilon)_h$, then we have $x \wedge 1 = 1 - (x - 1)^-$ by Definition 3.16. It follows from Definition 3.17 that

$$\tau(\varepsilon[1 - (x - 1)^-]) = \tau(\varepsilon[x \wedge 1]) \leq \tau(\varepsilon[x]).$$

Thus, similar to the proof of [12] Proposition 4.10, we have

$$\tau(\varepsilon[\lambda 1 - (x - \lambda 1)^-]) \leq \tau(\varepsilon[x]) \ for\ all\ \lambda > 0.$$

Now we regard x and 1 as elements in H_τ, so that x^- is the projection of x onto the closed convex cone $-H_\tau^+$. Using the continuity of the projection on Hilbert spaces and combining with the lower semicontinuity of Dirichlet form $\varepsilon_\tau(,)$ on H_τ, then it implies that

$$\tau(\varepsilon[x^-]) = \varepsilon_\tau[x^-] \le \liminf_{\lambda\to 0} \varepsilon_\tau[\lambda 1 - (x-\lambda 1)^-]$$

$$= \liminf_{\lambda\to 0} \tau(\varepsilon[\lambda 1 - (x-\lambda 1)^-]) \le \tau(\varepsilon[x]).$$

Replace x in the above equation with $\alpha x^+ - x^-$ for $\alpha > 0$, and note that $\varepsilon(,\ )$ is τ-real, so that

$$\overline{\tau(\varepsilon(x^+,\ x^-))} = \tau(\varepsilon(x^+,\ x^-)) = \tau(\varepsilon(x^-,\ x^+)).$$

Hence,

$$\tau(\varepsilon[x^-]) \le \tau(\varepsilon[\alpha x^+ - x^-]) = \tau(\varepsilon[x^-]) + \alpha^2\tau(\varepsilon[x^+]) - 2\alpha\tau(\varepsilon(x^+,\ x^-)),$$

from which it implies that $\tau(\varepsilon(x^+,\ x^-)) \le 0$ since α is any positive. Therefore,

$$\tau(\varepsilon[x]) = \tau(\varepsilon[x^+ - x^-]) = \tau(\varepsilon[x^+]) - 2\tau(\varepsilon(x^+,\ x^-)) + \tau(\varepsilon[x^-]) \ge \tau(\varepsilon[x^+]).$$

(2) Note that $|x| = x^+ + x^-$. By (1) we have

$$\tau(\varepsilon[|x|]) - \tau(\varepsilon[x]) = \tau(\varepsilon[x^+ + x^-]) - \tau(\varepsilon[x^+ - x^-]) = 4\tau(\varepsilon(x^+,\ x^-)) \le 0.$$

At the end of this subsection, Beurling-Deny type criterion is obtained under the above framework.

Theorem 3.13 *Let $T = \{T_t,\ 0 \le t < \infty\} \subseteq L(E)$ be a weak $*$ continuous, symmetric, and contractive regular module operator semigroup. Then T is Markovian if and only if $\varepsilon[x] =< Gx,\ x >\ \ (x \in D(G))$ is a B-valued Dirichlet form, where $-G$ is the infinitesimal generator of T.*

Proof Suppose that $\varepsilon[x] =< Gx,\ x >$ is a B-valued Dirichlet form. We now prove that $\{T_t\}_{t\ge 0}$ is a Markov module operator semigroup. By Proposition 3.8, it suffices to prove that each αG_α $(\alpha > 0)$ is a Markov module operator. Let $y \in E$ and $0 \le y \le 1$, write $x = \alpha G_\alpha y$, and we have to show that $0 \le x \le 1$.

Step 1. To prove that $x \ge 0$, and equivalently, to show that $x = x^+$., for this, set a B-valued function Φ as follows:

$$\Phi : D(\varepsilon) \to B^+,\ \Phi(z) = \alpha^{-1}\varepsilon[z] + < z - y,\ z - y >,$$

for all $z \in D(\varepsilon)$. In what follows, we shall prove that $x = \alpha G_\alpha y$ is the unique element of $D(\varepsilon)$ minimizing Φ, that is, to show that $\Phi(z) \ge \Phi(x)$ for all $z \in D(\varepsilon)$, and $\Phi(z) = \Phi(x) \Longleftrightarrow z = x$.

Indeed, observe that $x = \alpha G_\alpha y,\ \alpha G_\alpha = (I + \alpha^{-1}G)^{-1}$, where I is the unit of $L(E)$. We have

$$\Phi(z) - \Phi(x) = \alpha^{-1} < \sqrt{G}z,\ \sqrt{G}z > -\alpha^{-1} < \sqrt{G}x,\ \sqrt{G}x > + < z,\ z > -$$

$$< z,\ y > - < y,\ z > - < x,\ x > + < x,\ y > + < y,\ x >=$$

$$< (I + \alpha^{-1}G)z,\ z > - < (I + \alpha^{-1}G)x,\ x > -$$

$$< z,\ y > - < y,\ z > + < x,\ y > + < y,\ x >=$$

$$< (I + \alpha^{-1}G)(z - x),\ z > - < z - x,\ y >=$$

$$< (I + \alpha^{-1}G)(z - x),\ z - x > \geq 0\,. \tag{3.3}$$

It follows from the above (3.3) that $\Phi(z) \geq \Phi(x)$ for all $z \in D(\varepsilon)$. Moreover, we have $\Phi(z) = \Phi(x) \Longleftrightarrow z = x$. Thus, to prove $x = x^+$, it suffices to prove that $\Phi(x^+) \leq \Phi(x)$. Assume that $\Phi(x^+) > \Phi(x)$. Then it yields $\tau(\Phi(x^+) - \Phi(x)) > 0$. Since $\varepsilon[,\ ]$ is a Dirichlet form, $\tau(\varepsilon[x^+]) \leq \tau(\varepsilon[x])$ by Proposition 3.9. Hence,

$$\tau(\Phi(x^+)) = \alpha^{-1}\tau(\varepsilon[x^+]) + \tau(< x^+ - y, x^+ - y >) \leq$$

$$\alpha^{-1}\tau(\varepsilon[x]) + \tau(< x^+ - y,\ x^+ - y >).$$

Observe that

$$\tau(< x^+ - y,\ x^+ - y > - < x - y,\ x - y >) =$$

$$\tau(< x^+,\ x^+ > - < x^+,\ y >- < y,\ x^+ >$$

$$- < x,\ x > + < x,\ y > + < y,\ x >) =$$

$$\tau(< x^+,\ x^+ > - < x^+,\ y > - < y,\ x^+ > - < x^+,\ x^+ > - < x^-,\ x^- > +$$

$$< x^+,\ y > - < x^-,\ y > + < y,\ x^+ > - < y,\ x^- >) =$$

$$-\,\tau(< x^-,\ x^- > + < x^-,\ y > + < y,\ x^- >) \leq 0.$$

Hence, $\tau(\Phi(x^+)) \leq \tau(\Phi(x))$. This is in contradiction with the above hypothesis, so that $x = x^+$.

Step 2. We prove next that $x = x \wedge 1$. For this, it suffices to prove that $\tau(< x - x \wedge 1,\ x - x \wedge 1 >) = 0$, because τ is faithful. Since

$$\tau(< x - x \wedge 1,\ x - x \wedge 1 > + < x - x \wedge 1,\ x \wedge 1 - y >) =$$

$$\tau(< x - x \wedge 1,\ x - y >) = \tau(< x - x\wedge,\ (I - \alpha^{-1} G_\alpha^{-1})x >) =$$

$$-\alpha^{-1}\tau(< x - x \wedge 1,\ (G_\alpha^{-1} - \alpha I)x >) = -\alpha^{-1}\tau(\varepsilon(x - x \wedge 1,\ x)) =$$

$$-(2\alpha)^{-1}\tau(\varepsilon(x - x \wedge 1,\ x + x \wedge 1) + \varepsilon(x - x \wedge 1,\ x - x \wedge 1)) . \tag{3.4}$$

The the first term of (3.4)

$$\tau(\varepsilon(x - x \wedge 1,\ x + x \wedge 1) = \tau(\varepsilon[x] - \varepsilon[x \wedge 1] - \varepsilon(x \wedge 1,\ x) + \varepsilon(x,\ x \wedge 1)) =$$

$$\tau(\varepsilon[x] - \varepsilon[x \wedge 1]) - \tau(\varepsilon(x \wedge 1,\ x)) + \tau(\varepsilon(x,\ x \wedge 1)) =$$

$$\tau(\varepsilon[x]) - \tau(\varepsilon[x \wedge 1]) \geq 0,$$

in which we have used the τ-real symmetric property of the Dirichlet form $\varepsilon[,\ ]$. It is clear that the second term of (3.4) ≥ 0. Therefore,

$$\tau(< x - x \wedge 1,\ x - x \wedge 1 >) + \tau(< x - x \wedge 1,\ x \wedge 1 - y >) \leq 0.$$

Consequently,

$$\tau(< x - x \wedge 1,\ x \wedge 1 - y >) \leq -\tau(< x - x \wedge 1,\ x - x \wedge 1 >) \leq 0.$$

On the other hand,

$$\tau(< x - x \wedge 1,\ x \wedge 1 - y >) = \tau(< x - x \wedge 1,\ x \wedge 1 - 1 + 1 - y >) =$$

$$\tau(< x - x \wedge 1,\ x \wedge 1 - 1 >) + \tau(< x - x \wedge 1,\ 1 - y >) \geq$$

$$\tau(< x - x \wedge 1,\ x \wedge 1 - 1 >) = -\tau(< (1 - x)^-,\ (1 - x)^+ >) = 0.$$

Hence, combining the above proof, it implies that $\tau(< x - x \wedge 1,\ x \wedge 1 - y >) = 0$, from which yields $\tau(< x - x \wedge 1,\ x - x \wedge 1 >) = 0$, so that $x = x \wedge 1$.

Conversely, suppose that $\{T_t\}_{t \geq 0}$ is Markovian. We have to show that if $x \in D(\varepsilon)_h$, then $x \wedge 1 \in D(\varepsilon)_h$ and $\tau(\varepsilon[x \wedge 1]) \leq \tau(\varepsilon[x])$.

We first prove that if $x \in D(\varepsilon)_h$ and $\alpha \geq 0$, then $x \wedge \alpha 1 \in D(\varepsilon)_h$, and $\tau(\varepsilon(x - x \wedge \alpha 1,\ x \wedge \alpha 1)) \geq 0$. Indeed, since $x = (x - \alpha 1)^+ + x \wedge \alpha 1$, for proving $x \wedge \alpha 1 \in D(\varepsilon)_h$, it suffices to prove that $(x - \alpha 1)^+ \in D(\varepsilon)_h$. Set

$$\varepsilon^{(\beta)}(y,\ z) = \beta < y,\ z - \beta G_\beta z >,$$

for all $y,\ z \in E$, and for all $\beta > 0$. Then we have

$$\tau(\varepsilon^{(\beta)}((x-\alpha 1)^+,\ x \wedge \alpha 1)) = \tau(\beta < (x-\alpha 1)^+,\ x - \alpha 1 - \beta G_\beta(x \wedge \alpha 1) >)$$

$$= \tau(\beta < (x-\alpha 1)^+,\ x \wedge \alpha 1 >) - \tau(\beta < (x-\alpha 1)^+,\ \beta G_\beta(x \wedge \alpha 1) >)\ . \quad (3.5)$$

It is easy to check that the first term of the right-hand side of (3.5)

$$\tau(\beta < (x-\alpha 1)^+,\ x \wedge \alpha 1 >) \geq \alpha\beta\tau(< (x-\alpha 1)^+,\ 1 >),$$

and the second term of (3.5)

$$\tau(\beta < (x-\alpha 1)^+,\ \beta G_\beta(x \wedge \alpha 1) >) \leq \alpha\beta\tau(< (x-\alpha 1)^+,\ 1 >).$$

Therefore,

$$\tau(\varepsilon^{(\beta)}((x-\alpha 1)^+,\ x \wedge \alpha 1)) \geq 0.$$

Consequently,

$$\tau(\varepsilon^{(\beta)}((x-\alpha 1)^+,\ (x-\alpha 1)^+)) = \tau(\varepsilon^{(\beta)}((x-\alpha 1)^+,\ x - x \wedge \alpha 1)) =$$

$$\tau(^{(\beta)}((x-\alpha 1)^+,\ x)) - \tau(\varepsilon^{(\beta)}((x-\alpha 1)^+,\ x \wedge \alpha 1)) \leq$$

$$\tau(\varepsilon^{(\beta)}(x-\alpha 1)^+,\ x))\ . \quad (3.6)$$

Thus,

$$(3.6) \leq (\tau(\varepsilon^{(\beta)}[x]))^{1/2} \cdot (\tau(\varepsilon^{(\beta)}[(x-\alpha 1)^+]))^{1/2},$$

by Cauchy-Schwarz inequality. Hence,

$$\tau(\varepsilon^{(\beta)}[(x-\alpha 1)^+]) \leq \tau(\varepsilon^{(\beta)}[x]).$$

Let $\beta \to \infty$ in the above inequality. Then it implies that $\tau(\varepsilon[(x-\alpha 1)^+]) \leq \tau(\varepsilon[x])$. Hence, $(x-\alpha 1)^+ \in D(\varepsilon)_h$. Thus, $x^-,\ x^+,\ x \wedge 1 \in D(\varepsilon)_h$, whenever $\alpha = 0, 1$, respectively.

Finally, we have to prove that $\tau(\varepsilon[x \wedge 1]) \leq \tau(\varepsilon[x])$. Since

$$\tau(\varepsilon[x] - \varepsilon[x \wedge 1]) = \tau(\varepsilon(x - x \wedge 1,\ x + x \wedge 1)) =$$

$$\tau(\varepsilon(x - x \wedge 1,\ x - x \wedge 1)) + 2\tau(\varepsilon(x - x \wedge 1,\ x \wedge 1))\ , \quad (3.7)$$

in which we have used the τ-real symmetric property of the Dirichlet form in the second line of (3.7). It is clear that the first term of (3.7) is nonnegative. Next, in the inequality

$$\tau(\varepsilon^{(\beta)}(x - x \wedge 1, x \wedge \alpha 1)) = \tau(\varepsilon^{(\beta)}((x - \alpha 1)^+,\ x \wedge \alpha 1)) \geq 0,$$

obtained in the above proof process, let $\beta \to \infty$ and get $\tau(\varepsilon(x - x \wedge \alpha 1,\ x \wedge \alpha 1)) \geq 0$. Take $\alpha = 1$ in this inequality, it follows that the second term of (3.7) is nonnegative as well. It follows from the above proof that $\tau(\varepsilon[x \wedge 1]) \leq \tau(\varepsilon[x])$.

3.5 Hypercontractivity and Logarithmic Sobolev Inequality in Probability Gage Space

The study of hypercontractivity in bosonic system was pioneered by Nelson [26]. Then Gross [18] established the equivalence between the hypercontractivity of the semigroup $e^{-tL} : L^2(\mathbb{R}^d, \mu) \to L^2(\mathbb{R}^d, \mu)$ and the logarithmic Sobolev inequality verified by μ, where L is the Dirichlet form operator for the Gaussian measure μ on $\mathbb{R}^d$. The extension of Nelson's theorem to the fermonic case started with Gross' papers (see [19]). Namely, he adapted the argument in the bosonic case by considering a suitable Clifford algebra $\mathscr{C}(\mathbb{R}^d)$ on the fermion Fock space and noncommutative L^p space on this algebra after Segal [32]. In particular, hypercontractivity makes perfectly sense in this context by considering the corresponding Ornstein-Uhlenbeck semigroup:

$$U_t^{(-1)} = e^{-tN^{-1}} : L^2(\mathscr{C}(\mathbb{R}^d), \tau) \to L^2(\mathscr{C}(\mathbb{R}^d),\ \tau).$$

Here N^{-1} denotes the fermion number operator. After some partial results, the optimal time hypercontractivity bound in the fermionic case was finally obtained by Carlen and Lieb in [11]:

$$||U_t^{(-1)}||_{L^p \to L^r} = 1 \text{ if and only if } e^{-2t} \leq \frac{p-1}{r-1}.$$

As a by-product, Carlen and Lieb proved that the above hypercontractivity in the fermion case is equivalent to the following logarithmic Sobolev inequality :

$$\tau(|a|^2)\log|a|^2 - ||a||_2^2 \log ||a||_2^2 \leq 2 < a,\ N^{-1}a >,$$

for all $a \in \mathscr{C}(\mathbb{R}^d)$.

This line was continued by Biane [5] and Junge et al. [21], in which they extended Carlen and Lieb' work to q-Ornstein-Uhlenbeck semigroup $(-1 < q < 1)$ on q-Gaussian von Neumann algebra introduced by Bozèjko and Speicher [7], and free product semigroup of Ornstein-Uhlenbeck semigroup on free products of many

Clifford algebras, and then obtained optimal time estimates for hypercontractivity, respectively. But the relation between hypercontractivity and associated logarithmic Sobolev inequality was not considered in [5] and [21]. The abovementioned Gaussian von Neumann algebra is a type II_1-factor, and free products of many Clifford algebras generate a concrete finite von Neumann algebra.

In this section, we shall prove the equivalence between hypercontractivity and associated logarithmic Sobolev inequality of a class of quantum Markov semigroup and associated Dirichlet form based on a probability gage space, such that the above q-Ornstein-Uhlenbeck semigroup $(-1 < q < 1)$ and free product semigroup of Ornstein-Uhlenbeck semigroup are special cases. Furthermore, we shall describe the spectral gap and weak spectral gap properties of quantum Markov semigroup and the associated Dirichlet form in this noncommutative context.

3.5.1 Hypercontractivity and Logarithmic Sobolev Inequality

Definition 3.18 A Markov semigroup $\{T_t\}_{t\geq 0}$ in the interpolating family $L^p(A, \tau)$ is called hypercontractivity, if for every $1 < p < q < \infty$, there exist $a(p, q) \geq 0$ and $b(p, q) > 0$, such that for every $t > a(p, q)$, we have $||T_t x||_q \leq b(p, q)||x||_p$, for each $x \in L^p(A, \tau)$; it is called strict hypercontractivity if the constant $b(p, q) \leq 1$.

Remark 3.14 When $a > 0$ and $b \geq 0$, let $p(t) = 1 + (p-1)\, e^{2t/a}$. Then the above hypercontractivity in Definition 3.18 is equivalent to the following statement:

For all $p \in (0, \infty)$, we have

$$||T_t x||_{p(t)} \leq e^{b\,(\frac{1}{p} - \frac{1}{p(t)})}||x||_p,$$

for all $t \geq 0$ and $x \in L^p(A, \tau)$.

In the commutative case, it is equivalent to $||T_t x||_4 \leq c||x||_2$ for all $t \geq 0$; here c is a positive constant (see [30] Chapter 3, Theorem 3.2.2).

From now on, we are given a symmetric Markov semigroup $\{T_t\}_{t\geq 0} = \{e^{-Lt}\}_{t\geq 0}$ and its associated regular Dirichlet form $\varepsilon[x] =< \sqrt{L}x, \sqrt{L}x >$ on the above probability gage space (A, τ). Let $Ent(x) = \tau(x \log x) - ||x||_{L^2} \log ||x||_{L^2}$ denote the entropy of a positive element x.

Theorem 3.14 *Let a (> 0) and b (≥ 0) be constants. Set $p(t) = 1 + (p - 1)e^{2t/a}$, $b(t) = b(\frac{1}{p}) - \frac{1}{p(t)})$ whenever $p > 1$. Then the following statements are equivalent:*

(1) $||T_t x||_{p(t)} \leq e^{b(t)}||x||_p$ for all $x \in A$, $p > 1$ and $t \geq 0$;
(2) $Ent(|x|^2) \leq 4\, a\, \varepsilon[x] + b\, ||x||_2^2$ for all $x \in D(\varepsilon)$.

When $x \in D_h := (\varepsilon)D(\varepsilon) \cap L_h^2(A, \tau)$, the coefficient 4a in the inequality can be replaced by 2a.

We need the following lemma which plays a crucial role for proving Theorem 3.14.

Lemma 3.2 *For all invertible positive $x \in A \cap D(\varepsilon)$, and $1 < p < \infty$, one has*

$$\varepsilon[x^{p/2}] \leq \frac{p^2}{4(p-1)} \varepsilon\left(x,\ x^{p-1}\right).$$

That is,

$$< Lx^{p/2},\ x^{p/2} > \leq \frac{p^2}{4(p-1)} < Lx,\ x^{p-1} > .$$

Proof First, note that x is invertible and positive; then there exists a constant $c > 0$ such that Spec $(x) \subseteq [c,\ ||x||]$. Hence, by Proposition 3.6 (2) combining the function calculus of x, it is easy to check that $x^{\frac{p}{2}}$ and x^{p-1} are in $A \cap D(\varepsilon)$. By the spectrum decomposition of L, we have

$$< Lx^{p/2},\ x^{p/2} >= \lim_{t \to 0} \frac{1}{t} \tau\left[\left(x^{p/2} - T_t x^{p/2}\right) x^{p/2}\right];$$

$$< Lx,\ x^{p-1} >= \lim_{t \to 0} \frac{1}{t} \tau\left[(x - T_t x)\, x^{p-1}\right].$$

Next, it suffices to prove the following inequality:

$$\tau\left[\left(x^{p/2} - T_t x^{p/2}\right) x^{p/2}\right] \leq \frac{p^2}{4(p-1)} \tau\left[(x - T_t x)\, x^{p-1}\right]. \tag{3.8}$$

For any fixed $t > 0$. Since T_t is symmetric and Markovian, $\tau[\varphi(x)T_t(\psi(x))]$ is positive and linear in φ and ψ, for positive and continuous functions φ and ψ on $\mathbb{R}$, respectively. Moreover, when φ and ψ satisfy $\varphi(\alpha) \leq c|\alpha|$ and $\psi(\alpha) \leq c'|\alpha|$, where c, c' are constants, then by the property of normal traces on von Neumann algebras (see [2, 14]), there exists a positive measure μ_x on $\mathbb{R}\backslash\{0\} \times \mathbb{R}\backslash\{0\}$ with support contained in $\sigma(x) \times \sigma(x)$ such that $\mu_x(\alpha,\ \beta) = \mu_x(\beta,\ \alpha)$ and

$$\tau[\varphi(x)T_t(\psi(x))] = \int\int \varphi(\alpha)\psi(\beta)d\mu_x(\alpha, \beta).$$

Since $1 - T_t(1)$ is positive, there is a positive measure υ_x on $\mathbb{R}\backslash\{0\}$ with support contained in Spec(x) such that

$$\tau[\varphi(x)(1 - T_t(1))] = \int \varphi(\alpha)d\upsilon_x(\alpha).$$

Consider now the quadratic form $\tau[x(1 - T_t)x]$; we then have

$$\tau[\varphi(x)(1 - T_t)\varphi(x)] = \tau\left[\varphi(x)^2(1 - T_t(1))\right] + \tau\left[\varphi(x)^2\, T_t(1) - \varphi(x)T_t(\varphi(x))\right].$$

Therefore,

$$\tau[\varphi(x)(1 - T_t)\varphi(x)] = \int \varphi(\alpha)^2 d\nu_x(\alpha) + \frac{1}{2}\int\int [\varphi(\alpha) - \varphi(\beta)]^2\, d\mu_x(\alpha,\ \beta)\ . \tag{3.9}$$

In the following, let $\varphi(\alpha) = \alpha^{\frac{p}{2}},\ \alpha \in \sigma(x)$. Take $\varphi(x) = x^{\frac{p}{2}}$ in the above equation (3.9); it follows that

$$\tau[(x^{p/2} - T_t x^{p/2})x^{p/2}] = \int \alpha^p d\nu_x(\alpha) + \frac{1}{2}\int\int (\alpha^{p/2} - \beta^{p/2})^2 d\mu_x(\alpha,\ \beta).$$

Similarly, we can get

$$\tau[x^{p-1}(x - T_t x)] = \int \alpha^p d\nu_x(\alpha) + \frac{1}{2}\int\int (\alpha - \beta)(\alpha^{p-1} - \beta^{p-1} d\mu_x(\alpha,\ \beta)).$$

Hence, (3.8) holds true from the above two formulas combining the following fact:

$$(a^{p/2} - b^{p/2})^2 \le \frac{p^2}{4(p-1)}(a - b)(a^{p-1} - b^{p-1}),\ a,\ b \ge 0,\ p > 1.$$

The Proof of Theorem 3.14 From [28] Proposition 3.2, we see that the norm of T_t from $L_h^p(A,\ \tau)$ to $L_h^q(A, \tau)$ $(p,\ q > 1)$ is achieved on the positive cone $L_+^p(A,\ \tau)$. So it is sufficient to consider hypercontractivity on positive cones. Given an invertible positive $x \in A \cap D(\varepsilon)$, put $\varphi(t) = e^{-b(t)}||T_t x||_{p(t)}$. By [16] Lemma 2 and [23] Lemma 3.1, a straightforward calculus shows that

$$\begin{aligned}\frac{d}{dt}\log\varphi(t) &= \frac{d}{dt}\left(-b(t) + \log||T_t x||_{p(t)}\right)\\ &= -b'(t) + \frac{1}{||T_t x||_{p(t)}}\frac{d||T_t x||_{p(t)}}{dt}\\ &= -b'(t) + \frac{1}{||T_t x||_{p(t)}}\frac{d}{dt}[\tau(T_t x)^{p(t)}]^{\frac{1}{p(t)}}\ .\end{aligned} \tag{3.10}$$

Since

$$\frac{d}{dt}[\tau(T_tx)^{p(t)}]^{\frac{1}{p(t)}} = ||T_tx||_{p(t)}\frac{d}{dt}[\frac{1}{p(t)}\log\tau_q(T_tx)^{p(t)}]$$

$$= ||T_tx||_{p(t)}[-\frac{p'(t)}{p^2(t)}\log||T_tx||_{p(t)}^{p(t)} + \frac{1}{p(t)}\frac{1}{||T_tx||_{p(t)}^{p(t)}}\frac{d||T_tx||_{p(t)}^{p(t)}}{dt}],$$

and since

$$\frac{d||T_tx||_{p(t)}^{p(t)}}{dt} = \frac{d}{dt}\tau[(T_tx)^{p(t)}]$$

$$= \tau[(T_tx)^{p(t)}(p'(t)\log T_tx + p(t)(T_tx)^{-1}\frac{dT_tx}{dt})].$$

the above equation is brought into formula (3.10) to get the following equation:

$$\frac{d}{dt}\log\varphi(t) = -b'(t) - \frac{p'(t)}{p^2(t)}\log||T_tx||_{p(t)}^{p(t)}$$

$$+ \frac{1}{p(t)||T_tx||_{p(t)}^{p(t)}}\tau[(T_tx)^{p(t)}(p'(t)\log T_tx + p(t)(T_tx)^{-1}\frac{dT_tx}{dt})]$$

$$= -b'(t) - \frac{p'(t)}{p^2(t)}\log||T_tx||_{p(t)}^{p(t)} + \frac{1}{p(t)||T_tx||_{p(t)}^{p(t)}}\tau[p'(t)(T_tx)^{p(t)}\log T_tx]$$

$$+ \tau[p(t)(T_tx)^{p(t)-1}\frac{dT_tx}{dt}]\,. \tag{3.11}$$

Since $\frac{dT_tx}{dt} = -L(T_tx)$, we have

$$\tau[(T_tx)^{p(t)-1}\frac{dT_tx}{dt}] = -\varepsilon((T_tx)^{p(t)-1}, T_tx).$$

On the other hand, since

$$Ent((T_tx)^{p(t)}) = \tau[(T_tx)^{p(t)}\log(T_tx)^{p(t)}] - \tau[(T_tx)^{p(t)}\log\tau[(T_tx)^{p(t)}]]$$

$$= \tau[(T_tx)^{p(t)}\log(T_tx)^{p(t)}] - ||T_tx||_{p(t)}^{p(t)}\log||T_tx||_{p(t)}^{p(t)},$$

it follows combined with formula (3.11) that

$$\frac{d}{dt}\log\varphi(t) = -b'(t) + \frac{p'(t)}{p(t)^2}\frac{1}{||T_tx||_{p(t)}^{p(t)}} Ent((T_tx)^{p(t)})$$

$$-\frac{1}{||T_tx||_{p(t)}^{p(t)}}\varepsilon((T_tx)^{p(t)-1}, T_tx)\,. \tag{3.12}$$

Now we prove that (1) is equivalent to (2). First, if condition (1) is true, we want to prove that (2) is also true. Since $b(0) = 0$ and $p(0) = p$, $\varphi(0) = ||x||_p$. It follows from the hypercontractivity of T_t that $\varphi'(0) \leq 0$. Then, via the above formula (3.12), we have

$$Ent((T_tx)^{p(t)}) \leq \frac{p^2(t)||T_tx||_{p(t)}^{p(t)}}{p'(t)}[b'(t) + \frac{1}{||T_tx||_{p(t)}^{p(t)}}\varepsilon((T_tx)^{p(t)-1},\ T_tx)].$$

Let $t = 0$ and $p = 2$. It follows that $p(0) = 2$, $p'(0) = \frac{2}{a}$, $b'(0) = \frac{1}{2a}$. Therefore, from the above inequality, it implies that

$$Ent(x^2) \leq 2a\varepsilon[x] + b||x||_2^2.$$

Given any positive element $x \in D(\varepsilon)$. Since $\varepsilon[,]$ is regular, there exists a sequence (x_n) consisting of positive invertible elements in $A \cap D(\varepsilon)$ such that $|||x_n - x|||_1 \to 0$ as $n \to \infty$, where $|||x|||_1 = \varepsilon[x] + \tau(|x|^2)$ is the graph norm. It follows that $\varepsilon[x_n] \to \varepsilon[x]$ and $||x_n||_2 \to ||x||_2$, as $n \to \infty$. For any fixed $n \in \mathbb{N}$, by the above proof, we have

$$Ent(x_n^2) \leq 2a\varepsilon[x_n] + b||x_n||_2^2.$$

Letting $n \to \infty$, and combined with the continuity of norm, one can obtain

$$Ent(x^2) \leq 2a\varepsilon[x] + b||x||_2^2.$$

Hence, for any $y \in D(\varepsilon) \cap L_h^2(A,\ \tau)$, it results from the above inequality that

$$Ent(|y|^2) \leq 2a\varepsilon[|y|] + b||y||_2^2.$$

Since $\varepsilon[|y|] \leq \varepsilon[y]$ by Proposition 3.6, it follows combined with the above inequality that

$$Ent(|y|^2) \leq 2a\varepsilon[y] + b||y||_2^2.$$

Since $\varepsilon[|z|] \leq 2\,\varepsilon[z]$ for all $z \in D(\varepsilon)$ from Remark 3.10, it follows that

$$Ent(|z|^2) \leq 4\,a\varepsilon[z] + b||z||_2^2.$$

The above proof shows that condition (2) holds. Conversely, if condition (2) is true, it is proved that condition (1) is also true. For a given invertible positive $x \in A \cap D(\varepsilon)$, we have

$$Ent(x^2) \leq 2a\varepsilon[x] + b||x||_2^2.$$

Replacing x with $x^{\frac{p}{2}}$, one can get

$$Ent(x^p) \leq 2a\varepsilon[x^{\frac{p}{2}}] + b||x^{\frac{p}{2}}||_2^2.$$

It follows from Lemma 3.2 that

$$Ent(x^p) \leq \frac{ap^2}{2(p-1)}\varepsilon(x^{p-1}, x) + b||x||_p^p.$$

Take x be $T_t x$ and p be $p(t)$ in the above formula; then we get

$$Ent((T_t x)^{p(t)}) \leq \frac{ap(t)^2}{2(p(t)-1)}\varepsilon((T_t x))^{p-1}, T_t x) + b||T_t x||_{p(t)}^{p(t)}. \tag{3.13}$$

Therefore, from the above formula (12), it implies that

$$\frac{d}{dt}\log\varphi(t) = \frac{p'(t)}{p^2(t)||T_t x||_{p(t)}^{p(t)}}[Ent((T_t x)^{p(t)}) -$$

$$-\frac{p(t)^2}{p'(t)}\varepsilon((T_t x)^{p(t)-1}, T_t x) - \frac{b'(t)p(t)^2}{p'(t)}||T_t x||_{p(t)}^{p(t)}]. \tag{3.14}$$

Since $b(t) = b(\frac{1}{p} - \frac{1}{p(t)})$, $p(t) = 1 + (p-1)e^{\frac{2t}{a}}$, $b'(t) = \frac{p'(t)}{p(t)^2}$, $p'(t) = \frac{2}{a}(p-1)e^{\frac{2t}{a}}$, from which it implies that $\frac{p(t)^2}{p'(t)} = \frac{ap(t)^2}{2(p(t)-1)}$, and $\frac{b'(t)p(t)^2}{p'(t)} = b$. Compare (13) and (14), then we have $\frac{d}{dt}\log\varphi(t) \leq 0$; it follows that $\frac{d}{dt}\varphi(t) \leq 0$. Therefore, $\varphi(t) \leq \varphi(0) = ||x||_p$. Since $\varepsilon[,\ ]$ is regular, the subset of invertible positive elements in $A \cap D(\varepsilon)$ is dense in all $L^p(A,\ \tau)$ with respect to the L^p-norms for all $p > 1$, and since $\varphi(t)$ is continuous in the operator norm, from which it shows that $\{T_t\}$ has hypercontractivity. That is, condition (1) holds. Hence, condition (1) and condition (2) are equivalent.

If $\{T_t\}_{t\geq 0} = \{e^{-Lt}\}_{t\geq 0}$ is a completely positive Markov semigroup with associated completely and regular Dirichlet form $\varepsilon[x] =< Lx,\ x >$, based on the probability gage space A, then from Theorem 3.14, one can obtain the following consequence:

Corollary 3.1 *The notations are same as Theorem 3.14. Given a completely positive Markov semigroup* $\{T_t\}_{t\geq 0}$ *and its associated completely and regular Dirichlet form* $\varepsilon[\ ,\ ]$ *on* $L^2(A,\ \tau)$*. If* $\varepsilon[|J(x)|] = \varepsilon[|x|]$ *for all* $x \in D(\varepsilon)$*, then the following statements are equivalent:*

1. $||T_t x||_{p(t)} \leq e^{b(t)}||x||_p$ *for all* $x \in A,\ \forall p > 1,\ \forall t \geq 0$;
2. $Ent(|x|^2) \leq 2\ a\ \varepsilon[x] + b\ ||x||_2^2$ *for all* $x \in D(\varepsilon)$.

Proof Using Theorem 3.14 and the regularity of $\varepsilon[\ ,\ ]$, we only need to prove that $\varepsilon[|x|] \leq \varepsilon[x]$ for all $x \in A \cap D(\varepsilon)$. Indeed, since $\varepsilon[x] =< x,\ Lx >$ is completely Dirichlet form, $(\varepsilon^2,\ D(\varepsilon^2))$ is a Dirichlet form. Let $\tilde{x} = \begin{bmatrix} 0 & x^* \\ x & 0 \end{bmatrix}$. It is easy to see that $\tilde{x}$ is a self-adjoint element in $D(\varepsilon^2) := D(\varepsilon) \otimes M_2(\mathbb{C})$. By Proposition 3.6, we have $\varepsilon^2[|\tilde{x}|] \leq \varepsilon^2[\tilde{x}]$. Since $|\tilde{x}| = (\tilde{x}^*\tilde{x})^{1/2} = \begin{bmatrix} |x| & 0 \\ 0 & |x^*| \end{bmatrix}$. It follows by routine calculation that

$$\varepsilon[|x|] + \varepsilon[|x^*|] = \varepsilon^2[\begin{bmatrix} |x| & 0 \\ 0 & |x^*| \end{bmatrix}]$$

$$= \varepsilon^2[|\tilde{x}|] \leq \varepsilon^2[\tilde{x}] = \varepsilon^2[\begin{bmatrix} 0 & x^* \\ x & 0 \end{bmatrix}] = \varepsilon[x^*] + \varepsilon[x].$$

Therefore, from the above inequality, one can obtain $2\ \varepsilon[|x|] \leq 2\ \varepsilon[x]$ since $\varepsilon[x^*] = \overline{\varepsilon[x]} = \varepsilon[x]$ by the condition of this claim and $J(x) = x^*$ when $x \in A$). Hence, $\varepsilon[|x|] \leq \varepsilon[x]$.

Remark 3.15

(1) Theorem 3.14 and Corollary 3.1 are valid based on von Neumann algebra with a faithful normal semifinite trace.
(2) The condition $\varepsilon[|J(x)|] = \varepsilon[|x|]$ for all $x \in D(\varepsilon)$ in Corollary 3.1 is clearly valid in the commutative settings. This equation generally does not hold in the noncommutative cases. We next want to show that q-Ornstein-Uhlenbeck semigroup on q-Gaussian von Neumann algebra $(-1 < q < 1)$ introduced by Bozèjko and Speicher [7] satisfies this condition.

For this, we first recall briefly the construction of q-Gaussian von Neumann algebra and q-Ornstein-Uhlenbeck semigroup on it. Since the cases $q = \pm 1$ are well known, in the following, we shall only consider the cases for $-1 < q < 1$. For more details, we can refer to [5–9].

Suppose that $\mathscr{H}$ is an infinite-dimensional real separable Hilbert space with complexification $\mathscr{H}_{\mathscr{C}}$. Let Ω be a unit vector in a one-dimensional complex Hilbert space (disjoint from $\mathscr{H}_{\mathscr{C}}$). We refer to Ω as the vacuum, and by convention define $\mathscr{H}_{\mathscr{C}}^{\otimes 0} \equiv \mathbb{C}\Omega$. The algebraic Fock space $\mathscr{F}(\mathscr{H})$ is defined as

$$\mathscr{F}(\mathscr{H}) = \bigoplus_{n=0}^{\infty} \mathscr{H}_{\mathbb{C}}^{\otimes n},$$

where the direct sum and tensor product are algebraic. We then define a Hermitian form $< \cdot, \cdot >_q$ in $\mathscr{F}(\mathscr{H})$ as follows:

$$< \xi, \ \eta >_q = < \xi, \ P_q \eta >,$$

for all $\xi, \ \eta \in \mathscr{F}(\mathscr{H})$, in which $<, \ >$ is the usual scalar product, $P_q = \bigoplus_{n=0}^{\infty} P_q^{(n)}$, and $P_q^{(n)}(f_1 \otimes f_2 \otimes \cdots \otimes f_n) = \sum_{\pi \in S_n} q^{i(\pi)} f_{\pi(1)} \otimes f_{\pi(2)} \otimes \cdots \otimes f_{\pi(n)}$ for all $f_k \in \mathscr{F}(\mathscr{H}), \ k = 1, \ 2, \ \cdots, \ n, \ for \ all \ n \in \mathbb{N}$,

where S_n is the symmetric group on n symbols, and $i(\pi)$ counts the number of inversions in π, that is,

$$i(\pi) = \sharp\{(i, \ j) : 1 \leq i < j \leq k, \ \pi(i) > \pi(j)\}.$$

Indeed, the above Hermitian form is the conjugate-linear extension of

$$< \Omega, \ \Omega >_q = 1;$$

$$\begin{aligned} &< f_1 \otimes f_2 \otimes \cdots \otimes f_m, \ g_1 \otimes g_2 \otimes \cdots \otimes g_n >_q \\ &= \delta_{mn} \Sigma_{\pi \in S_n} q^{i(\pi)} < f_1, \ g_{\pi(1)} > \cdots < f_n, \ g_{\pi(n)} >, \end{aligned}$$

for $f_i, \ g_j \in \mathscr{H}, i = 1, \ 2, \ \cdots, \ m; \ j = 1, \ 2, \ \cdots, \ n; \ m, \ n \in \mathbb{N}$.

It is remarkable that, for $-1 < q < 1$, the form $< \cdot, \ \cdot >_q$ is always nondegenerate on $\mathscr{F}(\mathscr{H})$. The q-Fock space $\mathscr{F}_q(\mathscr{H})$ is defined as the completion of $\mathscr{F}(\mathscr{H})$ with respect to the inner product $< \cdot, \ \cdot >_q$.

For any vector $f \in \mathscr{H} \subset \mathscr{H}_{\mathbb{C}}$, define the creation operator $c_q(f)$ on $\mathscr{F}_q(\mathscr{H})$ to extend

$$c_q(f)\Omega = f$$

$$c_q(f) f_1 \otimes \cdots \otimes f_k = f \otimes f_1 \otimes \cdots \otimes f_k.$$

The annihilation operator $c_q^*(f)$ is its adjoint, which satisfies

$$c_q^*(f)\Omega = 0$$

$$c_q^*(f) f_1 \otimes \cdots \otimes f_k = \Sigma_{j=1}^{k} q^{j-1} < f_j, \ f > f_1 \otimes \cdots \otimes f_{j-1} \otimes f_{j+1} \otimes \cdots \otimes f_k.$$

The operators $c_q(f)$ and $c_q^*(f)$ are bounded on $\mathscr{F}_q(\mathscr{H})$ with norms as follows:

$$||c_q(f)|| = ||c_q^*(f)|| = \begin{cases} ||f||(1-q)^{-1/2}, & \text{if } 0 \le q < 1, \\ ||f||, & \text{if } -1 < q < 0. \end{cases}$$

Define the q-Gaussian von Neumann algebra $\Gamma_q(\mathscr{H})$ which is generated by the self-adjoint q-Gaussian operators $\omega(f) = c_q(f) + c_q^*(f)$, $f \in \mathscr{H}$ on $\mathscr{F}_q(\mathscr{H})$. It was proved in [8] Theorem 2.10 that $\Gamma_q(\mathscr{H})$ is a II_1-factor and $\tau_q(a) =< \Omega,\ a\Omega >_q$ for $a \in \Gamma_q(\mathscr{H})$, is the unique faithful normal trace on $\Gamma_q(\mathscr{H})$.

In the following, we shall introduce the concept of q-Ornstein-Uhlenbeck semigroup on the above q-Gaussian von Neumann algebra $\Gamma_q(\mathscr{H})$ for $-1 < q < 1$:

Let $T : \mathscr{H} \to \mathscr{H}$ be a contraction on the real Hilbert spaces $\mathscr{H}$ with complexification $T_{\mathbb{C}}$, and then the linear map defined on elementary tensors by

$$F_q(T)(f_1 \otimes \cdots \otimes f_n) = T_{\mathbb{C}} f_1 \otimes \cdots \otimes T_{\mathbb{C}} f_n$$

extends to a contraction

$$F_q(T) : \mathscr{F}_q(\mathscr{H}) \to \mathscr{F}_q(\mathscr{H}).$$

Now we define q-Gaussian functor Γ_q as a map

$$\Gamma_q(T) : \Gamma_q(\mathscr{H}) \to \Gamma_q(\mathscr{H})$$

in the following way:

1. $\Gamma_q(T)\omega(f) = \omega(Tf)$, for $f \in \mathscr{H}$;
2. $(\Gamma_q(T)(X))\Omega = F_q(T)(X\Omega)$.

By [8] Theorem 2.11 $\Gamma_q(T)$ is a unique bounded, normal, unital, completely positive, and trace-preserving map and is a covariant functor, that is, if T_1 and T_2 are contractions on $\mathscr{H}$, then

$$\Gamma_q(T_1 T_2) = \Gamma_q(T_1)\Gamma_q(T_2).$$

Furthermore, let $T_t = e^{-t} I_{\mathscr{H}}, t \ge 0$, where $I_{\mathscr{H}}$ is the identity on H. Then the q-Ornstein-Uhlenbeck semigroup is defined to be $\{U_t^{(q)}\}_{t\ge 0} = \{\Gamma_q(T_t)\}_{t\ge 0}$ on $\Gamma_q(\mathscr{H})$.

From the above construction, it is easy to see that $\{U_t^{(q)}\}_{t\ge 0}$ is a conservative, completely Markov semigroup on $\Gamma_q(\mathscr{H})$. From Remark 3.9, we see that $\{U_t^{(q)}\}_{t\ge 0}$ can be extended to completely Markov semigroup on all noncommutative $L^p(\Gamma_q(\mathscr{H}), \tau_q)$-spaces. Its generator N^q on $L^2(\Gamma_q(\mathscr{H}), \tau_q)$ is the number operator given by $N^q\Omega = 0$; and

$$N^q(f_1 \otimes \cdots \otimes f_n) = n(f_1 \otimes \cdots \otimes f_n),\ \ f_j \in \mathscr{H}_{\mathbb{C}}(j = 1,\ 2,\ \cdots, n).$$

Informally, we write $U_t^{(q)} = e^{-tN^q}$. From Theorem 3.12, we see that the corresponding quadratic form $\varepsilon[x] =< \sqrt{N^q}x, \sqrt{N^q}x >_q$ for all $x \in D(\sqrt{N^q})$, is a conservative, completely Dirichlet form.

P. Biane [5] described the strict hypercontractivity of the q-Ornstein-Uhlenbeck semigroup $\{U_t^{(q)}\}_{t\geq 0}$ and derived a logarithmic Sobolev inequality from strict hypercontractivity. However, the inverse proposition is not considered in [5]. In the end of this subsection, by using Corollary 3.1 we shall show that strict hypercontractivity and the logarithmic Sobolev inequality are equivalent for q-Ornstein-Uhlenbeck semigroup $\{U_t^{(q)}\}_{t\geq 0}$.

Theorem 3.15 *The notations are as in Theorem 3.14. Given the q-Ornstein-Uhlenbeck semigroup $\{U_t^{(q)}\}_{t\geq 0} = \{e^{-tN^q}\}_{t\geq 0}$ and its associated Dirichlet form $\varepsilon[x] =< N^q x, x >$ based on $\Gamma_q(\mathscr{H})$. Then the following statements are equivalent:*

(1) $||U_t^{(q)}x||_{p(t)} \leq ||x||_p$ for all $x \in \Gamma_q(\mathscr{H})$, $p > 1$ and $t \geq 0$;
(2) $Ent(|x|^2) \leq 2\,\varepsilon[x]$ for all $x \in D(\varepsilon)$.

Proof In order to prove the above claim, using Corollary 3.1 it suffices to prove the Dirichlet form $\varepsilon[x] =< N^q x, x >_q$, $x \in D(\varepsilon)$ is regular and satisfies the condition: $\varepsilon[|J(x)|] = \varepsilon[|x|]$ for all $x \in D(\varepsilon)$.

Indeed, given a fixed orthonormal basis $\{e_i\}_{i\in\mathbb{N}}$ in $\mathscr{H}$. For any subset $I = \{i_1, i_2, \cdots, i_n\} \subseteq \mathbb{N}$, by [8] Proposition 2.7, one can construct the q-Wick product $\psi_I = \psi(e_{i_1} \otimes e_{i_2} \otimes \cdots \otimes e_{i_n}) \in \Gamma_q(\mathscr{H})$ by induction as follows:

$$\psi(e_{i_k}) = \omega(e_{i_k}) = c_q(e_{i_k}) + c_q^*(e_{i_k});$$

$$\psi(e_{i_1} \otimes e_{i_2} \otimes \cdots \otimes e_{i_n}) = \omega(e_{i_1})\psi(e_{i_2} \otimes \cdots \otimes e_{i_n}) -$$

$$-\sum_{k=1}^{n} q^{k-1} < e_{i_1}, e_{i_k} > \psi(e_{i_1} \otimes \cdots \otimes \check{e_{i_k}} \otimes \cdots \otimes e_{i_n}),$$

where the symbol $\check{e_{i_k}}$ means that e_{i_k} has to be deleted in the product.

Define

$$\Phi : \Gamma_q(\mathscr{H}) \to \mathscr{F}_q(\mathscr{H})$$

through $\Phi(a) = a(\Omega)$. Since τ_q is a faithful trace, then this map is a continuous imbedding of $\Gamma_q(\mathscr{H})$ into $\mathscr{F}_q(\mathscr{H})$ which extends to an unitary isomorphism of $L^2(\Gamma_q(\mathscr{H}), \tau_q)$ with $\mathscr{F}_q(\mathscr{H})$. By routine calculation, we have

$$\Phi(\psi_I) = \psi_I \Omega = e_{i_1} \otimes e_{i_2} \otimes \cdots \otimes e_{i_n}.$$

Hence, from the construction of the q-Ornstein-Uhlenbeck semigroup $U_t^{(q)} = e^{-tN^q}$, we have $U_t^{(q)}\psi_I = e^{-tn}\psi_I$, it follows that $N^q\psi_I = n\psi_I$, which shows that $\psi_I \in D(\varepsilon)$. Denote by $\widetilde{\Gamma_q(\mathscr{H})}$ the set of all finite linear combinations of $\psi_I,\ I = \{i_1,\ i_2,\ \cdots,\ i_n\} \subseteq \mathbb{N}$. It follows that $\widetilde{\Gamma_q(\mathscr{H})} \subset D(\varepsilon)$ since $D(\varepsilon)$ is a subspace of $L^2(\Gamma_q(\mathscr{H}),\ \tau_q)$ and $\widetilde{\Gamma_q(\mathscr{H})}$ is dense in $D(\varepsilon)$ with respect to the graph norm. On the other hand, since $\widetilde{\Gamma_q(\mathscr{H})}$ is dense in $\Gamma_q(\mathscr{H})$ with respect to the operator norm, $A \cap D(\varepsilon)(\supset \widetilde{\Gamma_q(\mathscr{H})})$ is dense in $\Gamma_q(\mathscr{H})$ with operator norm. The above proof shows that the Dirichlet form $\varepsilon[,\ ]$ is regular.

Finally, in order to prove $\varepsilon[|J(x)|] = \varepsilon[|x|]$ for all $x \in D(\varepsilon)$, it is only needed to verify $\varepsilon[|(\psi_I)^*|] = \varepsilon[|\psi_I|]$ for all $\psi_I = \psi(e_{i_1} \otimes e_{i_2} \otimes \cdots \otimes e_{i_n}) \in \Gamma_q(\mathscr{H}) \cap D(\varepsilon)$. Since

$$< \psi_I\Omega,\ \psi_I\Omega >_q = \tau_q((\psi_I)^*\psi_I) = \tau_q(\psi_I(\psi_I)^*) = < (\psi_I)^*\Omega,\ (\psi_I)^*\Omega >_q,$$

there exists a unitary operator $U : L^2(\Gamma_q(\mathscr{H}),\ \tau_q) \to L^2(\Gamma_q(\mathscr{H}),\ \tau_q)$, such that $U|\psi_I| = |(\psi_I)^*|$; here we regard ψ_I and $\psi_I\Omega$ as the same. Furthermore, it is easy to check that $N^qU = UN^q$; hence,

$$\begin{aligned}\varepsilon[|(\psi_I)^*|] &= < N^q|(\psi_I)^*|,\ |(\psi_I)^*| >_q = < N^q(U|\psi_I|),\ U|\psi_I| >_q \\ &= < UN^q(|\psi_I|),\ U|\psi_I| >_q = < N^q(|\psi_I|,\ |\psi_I| >_q = \varepsilon[|\psi_I|].\end{aligned}$$

3.5.2 The Spectral Gap and Weak Spectral Gap Properties

This subsection is devoted to characterize the spectral gap and weak spectral gap properties of a class of Markov semigroup and its associated Dirichlet form based on a probability gage space.

It is well known that the exponential convergence of a Markov semigroup and spectral gap of its generator can be described by each other, and they are related to Poincaré inequality in the commutative real function space setting (see [15]). In order to describe L^2-convergence rates slower than exponential, M.Röckner and Feng-yu Wang [31] introduced weak Poincaré inequality, and they obtained the equivalence between a class of weak Poincaré inequality and the Kusuoka-Aida' "weak spectral gap property" (WSGP for short; see [1]) for conservative Dirichlet form as follows.

Proposition 3.10 (see [31] Proposition 1.2) *Suppose that* $(\varepsilon[f] = < Lf,\ f >$ *is a conservative Dirichlet form on real space* $L_h^2(X;\ \mu)$ *with respect to the probability space* $(X,\ \mathscr{B},\ \mu)$*, where* X *is a Hausdorff topological space, and* $\mathscr{B}$ *is the* σ*-algebra consisting of its Borel subsets with probability measure* μ *on* $(X, \mathscr{B})$*. Then the following statements are equivalent:*

(1) (a class of weak Poincaré inequality) $\mu(f^2) \le \alpha(r)\varepsilon[f] + r||f||_\infty^2,\ \mu(f) = 0,\ f \in D(L),\ r > 0$, *where* $\alpha(,)$ *is a nonnegative and decreasing function on* $(0, \infty)$;

(2) (WSGP) for any sequence $\{f_n\} \subset D(\varepsilon)$ *such that* $\mu(f_n^2) \le 1,\ \mu(f_n) = 0$, *and* $\varepsilon[f_n] \to 0$ *as* $n \to \infty$, *we have* $f_n \to 0$ *in probability.*

The focus of this subsection is to extend Proposition 3.10 to the noncommutative case.

Suppose that $\{T_t\}_{t\ge0} = \{e^{-tL}\}_{t\ge0}$ is a symmetric Markov semigroup based on a probability gage space $(A,\ \tau)$, and a regular Dirichlet form $\varepsilon[x] =< Lx,\ x >$ associated with it. If $\{T_t\}_{t\ge0}$ has strict hypercontractivity, then the following Poincaré inequality holds true:

Proposition 3.11 (Poincaré inequality) *Suppose that* $\{T_t\}_{t\ge0}$ *is a strict hypercontractivity symmetric Markov semigroup, and* $\varepsilon[x] =< Lx,\ x >,\ x \in D(\varepsilon)$ *is its associated regular Dirichlet form. Then*

$$\tau(|x|^2) - (\tau(|x|))^2 \le 2\ \varepsilon[x]\ for\ all\ x \in D(\varepsilon)\ (I).$$

When $x \in D(\varepsilon) \cap L_h^2(A,\ \tau)$, *the above inequality can be changed into the following form:*

$$\tau(x^2) - (\tau(x))^2 \le \varepsilon[x]\ (II).$$

The proof of Proposition 3.11 is similar to that in the commutative case (see [30], 3.2.3 Theorem (Spectral gap)), omitted here.

For example, the q-Ornstein-Uhlenbeck semigroup $\{U_t^{(q)}\}_{t\ge0}\ (-1 \le q \le 1)$ satisfies the above Poincaré inequality.

In what follows, we shall consider conservative, symmetric Markov semigroup $\{T_t\}_{t\ge0} = \{e^{-tL}\}_{t\ge0}$ and its associated regular Dirichlet form $\varepsilon[x] =< Lx,\ x >$, based on a probability gage space $(A,\ \tau)$. In this case, $T_t(1) = 1$, where 1 is the unit in A. Since the generator L of $\{e^{-tL}\}_{t\ge0}$ is a nonnegative, (unbounded) self-adjoint operator, so that its spectrum $\sigma(L) \subset [0,\ +\infty)$. Similar to the classical commutative cases, the spectral gap of L based on real noncommutative space $L_h^2(A,\ \tau)$ will be described. Since $L1 = 0$, then $0 \in \sigma(L)$ is an eigenvalue. Let $L_0^2 = \{x \in L_h^2(A,\ \tau) : \tau(x) = 0\}$, define $gap(L) := \inf\sigma(L|_{L_0^2})$, which is called the spectral gap of L, where $L|_{L_0^2}$ stands for the restriction of L onto L_0^2.

An equivalent characterization of $gap(L)$ is given as follows:

Theorem 3.16 $gap(L) = \inf\{\varepsilon[x] :\ \tau(x) = 0,\ \tau(x^2) = 1,\ x \in D_h(\varepsilon) := D(\varepsilon) \cap L_h^2(A,\ \tau)\}$.

Proof In order to prove this conclusion, it is equivalent to prove the following statement:

Given any fixed $\lambda \ge 0$. Then $gap(L) \ge \lambda$ if and only if $\varepsilon[x] \ge \lambda||x||_2^2$ for all $x \in L_0^2 \cap D_h(\varepsilon)$.

Indeed, if $gap(L) \geq \lambda$, then for all $x \in L_0^2 \cap D_h(\varepsilon)$, we have

$$T_t x = \int_0^\infty e^{-\alpha t} dE_\alpha x = \int_\lambda^\infty e^{-\alpha t} dE_\alpha x,$$

where $\{E_\alpha\}$ is the spectral resolution of L on L_0^2 (see [22] Theorem 5.2.2). Hence,

$$\varepsilon[x] =< Lx,\ x >= \lim_{t\to 0} < \frac{I - T_t}{t} x,\ x >$$

$$= \lim_{t\to 0} < \frac{\int_0^\infty dE_\alpha - \int_\lambda^\infty e^{-\alpha t} dE_\alpha}{t} x,\ x >$$

$$\geq \lim_{t\to 0} < \int_\lambda^\infty \frac{1 - e^{-\alpha t}}{t} dE_\alpha x,\ x >$$

$$=< \int_\lambda^\infty \lim_{t\to 0} \frac{1 - e^{-\alpha t}}{t} dE_\alpha x,\ x >\geq \lambda ||x||_2^2,$$

where I is the identity operator on L_0^2;

Conversely, assume that $\varepsilon[x] \geq \lambda ||x||_2^2$. Similar to the above proof, one can obtain

$$||T_t|_{L_0^2}|| \leq e^{-\lambda t},\ for\ all\ t \geq 0.$$

Therefore, $\{e^{\lambda t} T_t|_{L_0^2}\}$ is a C_0-contraction semigroup on L_0^2, and its infinitesimal generator is $\lambda - L$. Note that the spectrum set $\sigma(L) \subseteq [0, +\infty$; then by the Hille-Yosida Theorem (see [10], Theorem 3.1.10, or [29] Chapter 1, Theorem 3.1), $\sigma(\lambda - L) \subseteq (-\infty, 0)$, from which it implies that $gap(L) \geq \lambda$.

Remark 3.16 Similar to the proof of [15] Theorem 1.1.1, the following statements are equivalent:

(1) The above Poincaré inequality (II) holds true;
(2) $||T_t x||_2 \leq e^{-t} ||x||_2$ for all $x \in L_0^2$ and $t \geq 0$;
(3) $gap(L) \geq 1$.

Definition 3.19 (Ref. [27, 34]) Let $(A,\ \tau)$ be a probability gage space. Suppose that sequence $\{x_n\}$ is in A and x is a fixed element in A. We say $\{x_n\}$ converges to x in measure, if given any $\varepsilon > 0$, there exists a sequence $\{p_n\}$ of projections in A such that $||(x_n - x)p_n||_\infty < \varepsilon$, and $\tau(I - p_n) \to 0$ as $n \to \infty$, where I is the unit of A.

Theorem 3.17 *Let* $\varepsilon[x] =< Lx,\ x >$ *be a conservative, regular Dirichlet form on* $L^2(A,\ \tau)$, *where* $(A,\ \tau)$ *is a probability gage space. Then the following statements are equivalent:*

(1) *(Weak Poincaré inequality)* $||x||_2^2 \leq \alpha(\lambda)\varepsilon[x] + \lambda||x||_\infty$ *for* $x \in D_h(L)$ *with* $\tau(x) = 0$, *and* $\lambda > 0$, *where* $\alpha(\ ,\ )$ *is a nonnegative and decreasing function on* $(0,\ \infty)$;

(2) *(WSGP) for any sequence* $\{x_n\} \subset A \cap D_h(\varepsilon)$ *such that* $\tau(x_n^2) \leq 1,\ \tau(x_n) = 0$ *and* $\varepsilon[x_n] \to 0$ *as* $n \to \infty$, *then we have* $x_n \to 0$ *in measure.*

Proof Assume that (2) holds. If the weak Poincaré inequality in (1) is not established for any such $\alpha(\ ,\ )$. Note that $\varepsilon[\ ,\ ]$ is regular, so that there exists a $\lambda > 0$ and a sequence $\{x_n\} \subset A \cap D_h(\varepsilon)$ such that $\tau(x_n) = 0$, $\tau(x_n^2) = 1$ and

$$n\ \varepsilon[x_n] + \lambda||x_n||_\infty^2 < 1,\ n \geq 1.$$

From the above inequality, it implies that $||x_n||_\infty^2 < \lambda^{-1}$ for all n, and $\varepsilon[x_n] < \frac{1}{n} \to 0$ as $n \to \infty$. Using WSGP we have that for any given positive numberε, there exists a sequence $\{p_n\}$ of projections in A such that

$$||x_n p_n||_\infty < \varepsilon \text{ and } \tau(I - p_n) \to 0 \text{ as } n \to \infty.$$

Since

$$\begin{aligned} 1 = \tau(x_n^2) &= \tau(x_n^2 p_n + x_n^2(I - p_n)) \\ &= \tau(x_n^2 p_n) + \tau(x_n^2(I - p_n))\ , \end{aligned} \tag{3.16}$$

and

$$\tau(x_n^2 p_n) = \tau(x_n^2 p_n^2) = \tau((x_n p_n)^*(x_n p_n))$$

$$= ||x_n p_n||_2^2 \leq ||x_n p_n||_\infty^2 < \varepsilon^2,$$

$$\tau(x_n^2(I - p_n)) \leq (\tau(x_n^4))^{1/2}(\tau(I - p_n))^{1/2}$$

$$\leq ||x_n||_\infty^2(\tau(I - p_n))^{1/2} < \lambda^{-1}(\tau(I - p_n))^{1/2} \to 0 \text{ as } n \to \infty,$$

by Cauchy-Schwarz inequality. It follows combined with equation (3.16) that the above proof leads to the contradictory relation $1 \leq \varepsilon^2$ for any $\varepsilon > 0$.

Conversely, assume that (1) holds for some nonnegative and decreasing function $\alpha(\ ,\ )$ on $(0,\ \infty)$. Let $\{x_n\} \subset A \cap D_h(\varepsilon)$ with $\tau(x_n) = 0$, $\tau(x_n^2) \leq 1$ and $\varepsilon[x_n] \to 0$ as $n \to \infty$. By Definition 3.19, we have to prove that for any given $\varepsilon > 0$, there exists a nondecreasing sequence of projections $\{p_n\}$ in A such that $||x_n p_n||_\infty < \varepsilon$ and $\tau(I - p_n) \to 0$ as $n \to \infty$.

For any large $M > 0$, let $\chi_{[0,\ M^2]}$ be the characteristic function of the interval $[0,\ M^2]$. Set $p_M^n = \chi_{[0,\ M^2]}(x_n^2), n \in \mathbb{N}$, and we obtain a sequence p_M^n of projections in A. Therefore, from the inequality $M^2(I - p^n_M) \leq x_n^2$, it implies that $M^2\tau(I - p^n_M) \leq \tau(x_n^2) \leq 1$. Hence, $\tau(I - p^n_M) \leq M^{-2}$. Since

$$\tau(x_n p^n_M) + \tau(x_n(I - p^n_M)) = \tau(x_n) = 0,$$

it follows that

$$[\tau(x_n p^n_M)]^2 = [\tau(x_n(I - p^n_M))]^2 \leq \tau(x_n^2)\tau(I - p^n_M) \leq M^{-2}.$$

Since $x_n p^n_M = p^n_M x_n$, and $\tau(x_n p^n_M - \tau(x_n p^n_M)) = 0$, each $x_n p^n_M$ is a self-adjoint element. Using the weak Poincaré inequality for $x_n p^n_M - \tau(x_n p^n_M)$, by routine calculation, one can obtain

$$\tau(x_n p^n_M)^2 \leq (\tau(x_n p^n_M))^2 + 4\lambda M^2 + \alpha(\lambda)\varepsilon[x_n],\ \lambda > 0,\ n \geq 1,\ M > 0\,. \tag{3.17}$$

Furthermore, for any fixed $\varepsilon > 0$, $\varepsilon < M$, let $p^n_\varepsilon = \chi_{[0,\varepsilon^2]}(x_n^2)$. Since

$$\varepsilon^2(I - p^n_\varepsilon) = \varepsilon^2\chi_{(\varepsilon^2,+\infty)}(x_n^2) \leq x_n^2,$$

and since

$$p^n_\varepsilon p^n_M = p^n_M p^n_\varepsilon,\ x_n^2 p^n_M = p^n_M x_n^2,$$

from which it implies that

$$x_n^2 p^n_M \geq \varepsilon^2(I - p^n_\varepsilon)p^n_M.$$

Thus,

$$\tau(x_n p^n_M)^2 \geq \varepsilon^2\tau(p^n_M(I - p^n_\varepsilon)) = \varepsilon^2\tau(\chi_{(\varepsilon,M)}(|x_n|)). \tag{3.18}$$

Combining (3.18) with (3.17), we can get

$$\tau(\chi_{(\varepsilon,M)}(|x_n|)) \leq \varepsilon^{-2}[\alpha(r)\varepsilon[x_n] + M^{-2} + 4\lambda M^2],\ \lambda > 0,\ M > \varepsilon.$$

It follows from the above proof that $\tau(I - p^n_\varepsilon) \to 0$ as $n \to \infty$, since M and λ are arbitrary and they are independent with ε, and since $\varepsilon[x_n] \to 0$ as $n \to \infty$.

Finally, since $x_n^2 p^n_\varepsilon \leq \varepsilon^2 I$, $||x_n^2 p^n_\varepsilon||_\infty \leq \varepsilon^2$. Hence, $||x_n p^n_\varepsilon||^2_\infty = ||(x_n p^n_\varepsilon)^2||_\infty = ||x_n^2 p^n_\varepsilon||_\infty \leq \varepsilon^2$, that is, $||x_n p^n_\varepsilon||_\infty \leq \varepsilon$.

References

1. Aida, S.: Uniformly positivity improving property, Sobolev inequalities and spectral gap. J. Funct. Anal. **158**, 152–185 (1998)
2. Albeverio, S., Krohn, R.H.: Dirichlet forms and Markov semigroups on $C*$-algebras. Comm. Math. Phys. **56**, 173–187 (1977)
3. Arendt, W., Batty, C.J.K., Hieber, M., Neubrander, F.: Vector-valued Laplace Transforms and Cauchy Problems. Birkhäüser, Basel, Boston, Berlin (2001)
4. Beurling, A., Deny, J.: Dirichlet spaces. Proc. Natl. Acad. Sci. U.S.A. **45**, 208–215 (1959)
5. Biane, P.: Free hypercontractivity. Comm. Math. Phys. **184**, 457–474 (1997)
6. Bożejko, M., Speicher, R.: An example of a generalized Brownian motion. Commun. Math. Phys. **137**, 519–531 (1991)
7. Bożejko, M., Speicher, R.: Completely positive maps on Coxeter groups, deformed commutation relations, and operator spaces. Math. Ann. **300**, 97–120 (1994)
8. Bożejko, M., K*ü*mmerer, B., Speicher, R.: q-Gaussian processes: Non-commutative and classical aspects. Commun. Math. Phys. **185**, 129–154 (1997)
9. Bożejko, M.: Demonstratio Math. **XLV**(2), 129–154 (2012)
10. Bratteli, O., Robinson, D.W.: Operator algebras and quantum statistical mechanics I, C^*-and W^*-algebras. In: Symmetry Groups, Composition of States. Springer, Berlin/Heidelberg, New York (1979)
11. Carlen, E.A., Lieb, E.H.: Optimal hypercontractivity for Fermi fields and related noncommutative integration inequalities. Commun. Math. Phys. **155**, 27–46 (1993)
12. Cipriani, F.: Dirichlet forms and Markovian semigroups on standard forms of von Neumann algebras. J. Funct. Anal. **147**, 259–300 (1997)
13. Connes, A.: Noncommutative Geometry. Academic Press, New York (1994)
14. Davies, E.B., Lindsay, J.M.: Non-commutative symmetric Markov semigroups. Math. Z. **210**, 379–411 (1992)
15. Wang, F.: Functional Inequalities, Markov Semigroups and Spectral Theory. Mathematics Monograph Series 4. Science Press, Beijing (2005)
16. Fuglede, B., Kadison, R.V.: Determinant theory in finite factors. Ann. Math. **55**(3), 520–530 (1952)
17. Fukushima, M.: Dirichlet Forms and Markov Processes. North-Holland, Kodansha (1980)
18. Gross, L.: Hypercontractivity and logarithmic Sobolov inequalities for the Clifford Dirichlet forms. Duck Math. J. **42**(3), 383–396 (1975)
19. Gross, L.: Logarithmic Sobolev inequalities. Amer. J. Math. **97**, 1061–1083 (1975)
20. Guido, D., Isola, T., Scarlatti, S.: Non-symmetric Dirichlet forms on Semifinite von Neumann algebras. J. Func. Anal. **135**, 50–75 (1996)
21. Junge, M., Palazuelos, C., Parcbt, J., Perrin, M., Ricard, E.: Hypercontractivity for free products. Ann. Sci. Ecole. Norm. Sup. **48**(4), 861–889 (2015)
22. Kadison, R.V., Ringrose, J.R.: Fundamentals of the Theory of Operator Algebras I. In: Elementary Theory. Academic Press, New York (1983)
23. Kosaki, H.: Applications of uniform convexity of noncommutative L^p spaces. Trans. AMS **283**(1), 265–282 (1984)
24. Lance, E.C.: Hilbert C*-modules: A Toolkit for Operator Algebraists. London Mathematical Society, Lecture Note Series 124. Cambridge University Press, Cambridge (1994)
25. Magajna, B.: Hilbert C^*-modules in which all closed submodules are complemented. Proc. AMS **125**(3), 849–852 (1997)
26. Nelson, E.: A quartic interaction in two dimensions. In: Mathematical Theory of Elementary Particles, pp. 67–73. MIT Press, Cambridge, MA (1965)
27. Nelson, E.: Notes on non-commutative integration. J. Funct. Anal. **15**, 103–116 (1974)
28. Olkiewicz, R., Zegarlinski, B.: Hypercontractivity in noncommutative L^p spaces. J. Funct. Anal. **161**(1), 246–285 (1999)

29. Pazy, A.: Semigroups of Linear Operators and Applications to Partial Diffential Equations. Springer, Berlin/Heidelberg, New York (1983)
30. Accardi, L., Fagnola, F. (eds.): Quantum Interacting Particle Systems, QP-PQ, Quantum Probability and White Noise Analysis, vol. XIV. World scientific, New Jersey, London, Singapore, Hong Kong (2002)
31. Röckner, M., Wang, F.-Y.: Weak Poincaré inequalities and L^2-convergence rates of Markov semigroups. J. Funct. Anal. **185**, 564–603 (2001)
32. Segal, I.E.: A non-commutative extension of abstract integration. Ann. Math. **57**, 401–456 (1953)
33. Silverstein, M.L.: Symmetric Markov Processes. Lecture Notes in Mathematics, vol. 426. Springer, Berlin/Heidelberg, New York (1974)
34. Stinespring, W.F.: Integration theorems for gages and duality for unimodular groups. Trans. Amer. Soc. **90**, 15–56 (1959)
35. Toda, M., Kubo, R., Saitô, N.: Statistical Physics, I, II. Springer, Berlin/Heidelberg, New York (1992)
36. Lunchuan, Z.: Equivalence between logarithmic Sobolev inequality and hypercontractivity in a probability gage space. Proc. Edinb. Math. Soc. **64**(1), 59–71 (2021)

Appendix A
Fundamentals of C^*-Algebras and von Neumann Algebras

A.1 Basic Concepts of C^*-Algebras

Definition A.1 An algebra A (over $\mathbb{R}$ or $\mathbb{C}$) is said to be a normed algebra when A is a normed space such that

$$||ab|| \leq ||a|| \cdot ||b||,$$

for all $a\ b \in A$. If A is a Banach space relative to this norm, then it is said to be a Banach algebra. That is, a complete normed algebra is called a Banach algebra.

Definition A.2 If A is a Banach algebra, and if there is a $*$ operation on A (i.e., involution operation $a \to a^*$) such that

$$||a^*a|| = ||a||^2,$$

for all a in A. Then A is called a C^*-algebra, in which the equation is called the C^*-identity.

If C^*-algebra A has unit 1, then $||1|| = 1$, where 1 on the right is the number 1. Indeed, from the C^*-identity, $||1|| = ||1^*1|| = ||1||^2$.

A C^*-algebra A is commutative or abelian if it satisfies that $ab = ba$ for all $a,\ b$ in A.

Example A.1 The complex field $\mathbb{C}$ is a unital C^*-algebra, in which $*$ operation is the conjugate operation of complex number given by $\lambda \to \bar{\lambda}$ $(\lambda \in \mathbb{C})$.

Example A.2 If X is a locally compact Hausdorff space. Denote by $C_0(X)$ the set of all continuous functions from X to $\mathbb{C}$ vanish at infinity ($f \in C_0(X)$ means that if for each positive number ε then the set $\{x \in X | |f(x)| \geq \varepsilon\}$ is compact in X). Then $C_0(X)$ is a commutative C^*-algebra with involution $f \to \bar{f}$ $(f \in C_0(X))$. We shall see later that any commutative C^*-algebra is $*$ isomorphic to some $C_0(X)$.

L. Zhang, *Hilbert C^*- Modules and Quantum Markov Semigroups*,
https://doi.org/10.1007/978-981-99-8668-2

Example A.3 Let $M_n(\mathbb{C})$ be the set of all $n \times n$ matrices over the complex field $\mathbb{C}$; then it is an algebra with matrix multiplication and can become a Banach algebra when equipped with a norm. Moreover, $M_n(\mathbb{C})$ is a C^*-algebra with the conjugate transpose operation as $*$ operation given by

$$* : (a_{ij}) \to (\overline{a_{ji}}) \ ((a_{ij}) \in M_n(\mathbb{C})).$$

In general, matrix multiplication does not satisfy commutative law, so $M_n(\mathbb{C})$ is a noncommutative algebra.

Example A.4 Let H be a Hilbert space. Then $B(H)$ of all bounded linear operators is a noncommutative C^*-algebra with respect to the operator norm and conjugate of operators.

Definition A.3 Let A be a C^*-algebra and a and b in A. If $a = a^*$, then a is called self-adjoint or Hermitian; if $a^*a = aa^*$, then a is called a normal; moreover, if A has unit 1 and if $a^*a = aa^* = 1$, then a is called unitary. The set of all Hermitian elements in A is denoted by A_{sa}.

Suppose that A is a unital C^*-algebra with unit 1. An element a in A is said to be invertible if there exists b in A such that $ab = ba = 1$. In this case, the b is uniquely determined by a and is rewritten as a^{-1}. Denote by $Inv(a)$ the set of all invertible elements in a unital C^*-algebra A; then it is a multiplication group with respect to the multiplication operations in A.

Definition A.4 Let A be a unital C^*-algebra, and let an element a be in A. Then the spectrum of a is to be the following set:

$$\sigma(a) = \{\lambda \in \mathbb{C} | \lambda 1 - a \notin Inv(A)\}.$$

It can be proved that $\sigma(a)$ is a non-empty compact subset in $\mathbb{C}$.

If A is not unital, let $\tilde{A} = A \oplus \mathbb{C}$. In this case, $\sigma(a) = \{\lambda \in \mathbb{C} | \lambda 1 - a \notin Inv(\tilde{A})\}$.

Example A.5 If X is a compact Hausdorff space, then $C(X)$ is a unital abelian C^*-algebra. It is easy to check for any $f \in C(X)$, $\sigma(f) = \{f(x) | x \in X\}$.

Example A.6 If A is $M_n(\mathbb{C})$ and $a \in A$, then $\sigma(a)$ is the set of eigenvalues of a.

Thus, the spectrum of an element in C^*-algebra can be regarded as simultaneously a generalization of the range of a function and the set of eigenvalues of a finite square matrix.

Definition A.5 Let A be a C^*-algebra and a nonempty subset S in A. Set $S^* = \{a^* | a \in S\}$. If $S^* = S$, then S is called self-adjoint subset of A. In addition, a self-adjoint closed subalgebra of A is itself a C^*-algebra, which is called C^*-subalgebra of A.

Definition A.6 Let A be a C^*-algebra and let B be a C^*-subalgebra in A. If $A \cdot B \subset B$, then B is called a close left ideal of A. If $B \cdot A \subset B$, then B is called a closed right ideal of A. If B is simultaneously a closed left ideal and a closed right ideal, then it is called a closed ideal of A, or ideal for short. If A contains only two trivial ideals of $\{0\}$ and A itself, then A is called a simple C^*-algebra. Simple C^*- algebras are the cornerstone of constructing complex C^*-algebras.

From the GNS construction mentioned later, each C^*-algebra can be regarded as a C^*-subalgebras of $B(H)$ in the sense of $*$-isomorphism.

A.2 Gelfand Representation of Commutative C^*-Algebras and Functional Calculus of Normal Elements of C^*-Algebras

A.2.1 Gelfand Representation of Commutative C^*-Algebras

Definition A.7 Let A and B be C^*-algebras. If a homomorphic mapping $\varphi : A \to B$ preserves $*$ operation given by $\varphi(a^*) = (\varphi(a))^*$ $(a \in A)$, then it is called a $*$-homomorphism. Moreover, φ is said to be an isomorphism from A to B if it is injective, which is denoted by $A \cong B$.

Let A be a commutative C^*-algebra. Set $\Omega(A)$ as the character space of A, that is, the set of nonzero $*$-homomorphisms from A to $\mathbb{C}$. It is easy to check that $\Omega(A)$ is a locally compact Hausdorff space. If A is unital, then $\Omega(A)$ is compact.

With a in A. Define a function $\hat{a}$ from $\Omega(A)$ to $\mathbb{C}$ given by $\hat{a}(\tau) = \tau(a)$. We endow $\Omega(A)$ with the smallest vector topology such that all $\hat{a}$ $(a \in A)$ are continuous, which is called weak*-topology. Hence, the set $\{\tau \in \Omega(A) : |\tau(a)| \geq \varepsilon\}$ is weak * closed in the closed unit ball of A^* for each $\varepsilon > 0$, and weak * compact by the Banach-Alaoglu theorem. Thus, $\hat{a} \in C_0(\Omega(A))$. $\hat{a}$ is said to be the Gelfand transform of a.

Theorem A.1 *(Gelfand representation theorem) If A is a commutative C^*-algebra, then the Gelfand transform*

$$\varphi : A \to C_0(\Omega(A)) \ (a \to \hat{a})$$

is an isometric $*$ isomorphism from A to $C_0(\Omega(A))$.

A.2.2 Function Calculus of Normal Elements of C^*-Algebras

Let A be a C^*-algebra and a non-empty subset S in A. We write $C^*(S)$ for the C^*-subalgebra in a generated by S. $C^(S)$ is the smallest C^*-subalgebra of A containing S, which equals the intersection of all C^*-subalgebras of A containing

S. In particular, we denote by $C^*(a)$ if S is a single point set $\{a\}$. If a is normal, that is, $a^*a = aa^*$, then $C^*(a)$ is abelian.

Proposition A.1 *Suppose that A is a unital C^*-algebra and a normal element a in A. Then there exists a unique isomorphism φ from $C(\sigma(a))$ to $C(a, 1)$ of A generated by a and the unit 1, such that $\varphi(1) = 1$ and $\varphi(j) = a$, where j denotes the function $j(\alpha) = \alpha$ $(\alpha \in \sigma(a))$.*

For any continuous function f defined on $\sigma(a)$, write $f(a)$ for the image of f under the above isomorphism φ. Thus, $f(a) \in C^*(a, 1) \subset A$ and following basic properties are obtained.

Proposition A.2 *For all $f, g \in C(\sigma(a))$ and $\lambda, \mu \in \mathbb{C}$, the following statements hold:*

(1) $(\lambda f + \mu g)(a) = \lambda f(a) + \mu g(a)$;
(2) $(fg)(a) = f(a)g(a)$;
(3) $\bar{f}(a) = f(a)^*$;
(4) $\sigma(f(a)) = \{f(\lambda) | \lambda \in \sigma(a)\}$;
(5) $||f(a)|| = sup\{|f(\lambda)| | \lambda \in \sigma(a)\}$.

Moreover, if $g \in C(f(\sigma(a)))$, then $(g \circ f)(a) = g(f(a))$. In addition, if A has no unit, consider $\tilde{A} = A \oplus \mathbb{C}$. In this case, $\sigma_A(a) = \sigma_{\tilde{A}} \setminus \{0\}$. Therefore, in the above properties, it is required that $f(0) = 0$.

A.3 Positive Cone and Ordered Structure in C^*-Algebra

We consider $A = C(\Omega)$, where Ω is a compact Hausdorff space. Recall that if $a \in A$, then $\sigma(a) = \{a(\omega) : \omega \in \Omega\}$. Hence function a in A is positive if and only if $\sigma(a) \subset \mathbb{R}^+$. Moreover, it is easy to check that $a \geq 0$ is equivalent to the existence of $a' \in A$ such that $a = \bar{a'}a'$, and is also equivalent to $||a - \lambda 1|| \leq \lambda$ for any $\lambda \in \mathbb{R}$ and $\lambda \geq ||a||$. In this case, there is a unique positive square root $a^{\frac{1}{2}}$ of a such that $a = (a^{\frac{1}{2}})^2$. We denote by A^+ the set of all positive elements in A. Then A^+ is a positive cone and induces the partial order relation on A_{sa} consisting of all real-valued functions in A, that is, given $a, b \in A_{sa}$, then we have $a \leq b$ if and only if $b - a \in A^+$.

The above results can be generalized to any C^*-algebras.

Definition A.8 Let A be a C^*- algebra. Then an element a in A is said to be positive if $\sigma(a) \subset \mathbb{R}^+$. Denote by A^+ the set of all positive elements in A. In particular, if A is $B(H)$, then $T \in B(H)^+ \Leftrightarrow < Tx, x > \geq 0$ for all $x \in H$.

Proposition A.3 *Suppose that A is a C^*-algebra and a self-adjoint element a in A. Then $a \in A^+ \Leftrightarrow ||a - \lambda 1|| \leq \lambda$ for all $\lambda \in \mathbb{R}$ and $\lambda \geq ||a||$.*

From Definition A.8, we see that A^+ is a closed subset of A. By Proposition A.3 and combined with Definition A.8, it is easy to check that A^+ has the following basic properties:

(1) $\lambda\, a \in A^+$ if $a \in A^+$ and $\lambda \in \mathbb{R}^+$;
(2) $a + b \in A^+$ if a and b in A^+;
(3) $ab \in A^+$ if $a, b \in A^+$, and $ab = ba$;
(4) If $a \in A^+$ and $-a \in A^+$, then $a = 0$.

Hence, A^+ is a cone, which is called the positive cone of A.

The following is a series of equivalent characterizations of the positive elements of C^*-algebras.

Proposition A.4 *Let A be a C^*-algebra and an element a in A. Then the following conditions are equivalent:*

(1) $a \in A^+$;
(2) There exists a $b \in A^+$ such that $a = b^2$;
(3) There exists a $b \in A$ such that $a = b^*b$.

Remark A.1 (1) $A^+ = \{a^*a | a \in A\}$;
(2) It can be proved that the above $b \in A^+$ is unique, which is called the square root of a. At this time, we rewrite b as $a^{\frac{1}{2}}$.
(3) If a in A is a self-adjoint element, then there exists only $a^+,\ a^- \in A$ such that $a^+a^- = a^-a^+ = 0$ and $a = a^+ - a^-$.

A partial order relation is introduced on A_{sa} by the following way:

$$a, b \in A_{sa},\ a \leq b \Leftrightarrow b - a \in A^+.$$

The partial order relation has the following basic properties:

Proposition A.5 *Suppose that A is a C^*-algebra. Then the following statements hold:*

(1) If $a,\ b \in A_{sa}$, then $c^*ac \leq c^*bc$ from $a \leq b$, for each fixed $c \in A$;
(2) If $0 \leq a \leq b$, then $||a|| \leq ||b||$;
(3) If A is unital and $a,\ b$ are positive invertible elements in A, then $0 \leq b^{-1} \leq a^{-1}$ from $a \leq b$;
(4) If $a,\ b \in A^+$, then $a^{\frac{1}{2}} \leq b^{\frac{1}{2}}$ from $a \leq b$.

Remark A.2 $0 \leq a \leq b \Rightarrow a^2 \leq b^2 \Leftrightarrow A$ is an abelian C^*-algebra.

With the order relation, we can introduce some important concepts such as approximation unit and hereditary C^*-subalgebras. The concept of approximation unit plays a fundamental and important role in the study of C^* algebras.

Definition A.9 Let A be a C^*-algebra. Then an increasing net $(u_\alpha)_{\alpha \in \Gamma}$ of positive elements in the closed unit ball of A is called an approximate unit of A if it satisfies the condition that $a = \lim_\alpha a u_\alpha$ for all $a \in A$.

It can be proved that each C^*-algebra has an approximate unit.

Definition A.10 A C^*-subalgebra B of C^*-algebra A is said to be hereditary if for $a \in A^+$ and $b \in B^+$ the inequality $a \leq b$ implies $a \in B$.

It is easy to check that B is a hereditary C^*-subalgebra of A if and only if $BAB \subset B$. It is clear that closed two-sided ideals of A are all hereditary C^*-subalgebras of A.

Let $\mathfrak{L}$ be the set of all closed left ideals of A, and let $\mathfrak{B}$ be the set of all Hereditary C^*-subalgebras of A. Then there exists a bijection Φ from $\mathfrak{L}$ to $\mathfrak{B}$ given by

$$\Phi : \mathfrak{L} \to \mathfrak{B}, L \to L \cap L^* \ (L \in \mathfrak{L}).$$

Example A.7 If A is a C^*-algebra and a is in A^+, then the norm closure $\overline{(aAa)}$ of aAa is a hereditary C^*-subalgebra of A. In particular, if p is a projection in a C^*-algebra A, then the C^*-aubalgebra pAp is hereditary.

In the separable case, every hereditary C^*-subalgebra is of the form in the preceding example.

Proposition A.6 *Suppose that B is a nontrivial C^*-subalgebra in C^*-algebra A. Then B is hereditary if and only for each state φ on B there is a unique state $\overline{\varphi}$ on A extending φ.*

A.4 Positive Linear Functionals and GNS Construction

Definition A.11 Let A be a C^*-algebra and a linear functional $\rho : A \to \mathbb{C}$. Then ρ is said to be Hermitian if $\rho(a^*) = \overline{\rho(a)}$ $(a \in A)$. ρ is said to be positive provided $\rho(a) \geq 0$ when $a \in A^+$.

It is easy to show that positive linear functional is always Hermitian and bounded. ρ is Hermitian if and only if $\rho(a) = \overline{\rho(a)}$ for all $a \in A_{sa}$.

The following is a series of equivalent characterizations of positive linear functionals.

Proposition A.7 *Suppose that ρ is a bounded linear functional on C^*-algebra A. Then the following conditions are equivalent:*

(1) ρ is positive
(2) For each approximate unit $(u_\lambda)_{\lambda \in \Lambda} \subset A$, $||\rho|| = \lim_\lambda \rho(u_\lambda)$;
(3) For some approximate unit $(u_\lambda)_{\lambda \in \Lambda} \subset A$, $||\rho|| = \lim_\lambda \rho(u_\lambda)$.

Moreover, if A has unit 1, then the above conditions are also equivalent to $\rho(1) = ||\rho||$.

Proposition A.8 *Suppose that A is a C^*-algebra and ρ is a positive linear functional on A. Then*

$$|\rho(b^*a)|^2 \leq \rho(a^*a)\ \rho(b^*b)\ (a,\ b \in A).$$

Definition A.12 A positive linear functional ρ on a given C^*-algebra A is called a state on A if $||\rho|| = 1$. The set $S(A)$ of all states on A is called the state space of A. It can be proved that $S(A)$ is a locally compact convex subset in A^*. If A is unital, then $S(A)$ is a compact convex subset in A^*.

Given a Hilbert space H, for any $x \in H$ and $||x|| = 1$, then $\omega_x(T) =< Tx,\ x >$ $(T \in B(H))$ is a state on $B(H)$, which is called the vector state.

Example A.8 Let $A = M_n(\mathbb{C})$, and then the trace functional

$$tr : A \to \mathbb{C},\ (\lambda_{ij}) \to \sum_{i=1}^{n} \lambda_{ii}$$

is a positive linear functional.

Definition A.13 A representation of a C^*-algebra A is a pair $(H,\ \varphi)$, where $\varphi : A \to B(H)$ is a $*$-homomorphism and H is a Hilbert space. It is said to be faithful if φ is injective.

If a representation (H, φ) of A admits a vector ξ such that $[\varphi(A)\xi] = H$, where $[\varphi(A)\xi]$ denotes the closed subspace of H generated linearly by $\varphi(A)\xi$, then $(H,\ \varphi)$ is said to be cyclic and ξ is called a cyclic vector for φ.

Example A.9 Suppose that $(\Omega,\ \mathfrak{B},\ m)$ is a σ-finite measure space with a σ-algebra $\mathfrak{B}$ and a σ-finite measure m on it. Let $A = L_\infty(\Omega,\ \mathfrak{B}, m)$ be the abelian C^*-algebra consisting of (classes of) essentially bounded complex-valued measurable functions on Ω, and let $H = L_2(\Omega,\ \mathfrak{B},\ m)$ be the Hilbert space consisting of square integrable complex-valued functions on Ω. Given any fixed $f \in L_\infty(\Omega,\ \mathfrak{B}, m)$, define multiplication operator M_f on H given by

$$(M_f x)(\omega) = f(\omega)x(\omega)\ (\omega \in \Omega,\ x \in L_2(\Omega, \mathfrak{B},\ m)).$$

Then M_f is in $B(H)$ for all $f \in A$, and therefore it is easy to check that the mapping

$$\varphi : A \to B(H),\ f \to M_f\ (f \in L_\infty(\Omega,\ \mathfrak{B},\ m))$$

is a faithful $*$-representation of A.

Theorem A.2 *(GNS construction) If ρ is a state of a C^*-algebra A, then there is a cyclic representation π_ρ of A on a Hilbert space H_ρ, and a unit cyclic vector*

$x_\rho \in H_\rho$ such that

$$\rho(a) =< \pi_\rho(a)x_\rho,\ x_\rho > \quad (a \in A).$$

It can be further obtained the following Gelfand-Neumark theorem.

Theorem A.3 *(Gelfand-Neumark) Every C^*-algebra has a faithful $*$-representation.*

A.5 Basic Knowledge of Completely Positive Mappings

Suppose that A is a C^*-algebra. Then we can construct $n \times n$-matrix C^*-algebra $M_n(A)$ associated with A, for each $n \in \mathbb{N}$.

Proposition A.9 *An element a in $M_n(A)$ is positive if and only if there exist $a_1^{(k)}, \cdots, a_n^{(k)} \in A$ $(k = 1, \cdots, p)$ such that $a = \sum_{k=1}^{p} c^{(k)}$, where the matrix $c^{(k)} = (a_{ij}^{(k)})_{n\times n}$ is in $M_n(A)$, and $a_{ij}^{k} = a_i^{(k)*} a_j^{(k)}$.*

Proof In order to prove the sufficiency, it suffices to prove that matrix of form $(a_i^* a_j)_{n\times n}$ $(a_1, \cdots, a_n \in A)$ is positive in $M_n(A)$. Indeed, let $a_{1j} = a_j,\ 1 \le j \le n$, and $a_{ij} = 0,\ 2 \le i \le n,\ 1 \le j \le n$. Since $(a_i^* a_j)_{n\times n} = (a_{ij})^*(a_{ij})$, it follows that $(a_i^* a_j)_{n\times n}$ is positive in $M_n(A)$. We next prove the necessity. If $a = (a_{ij})$ is positive in $M_n(A)$, then there exists a matrix $b = (b_{ij}) \in M_n(A)$ such that $a = b^*b$, from which it implies that $a_{ij} = \sum_{k=1}^{n} b_{ki}^* b_{kj}$. Let $a_i^{(k)} = b_{k,i},\ i = 1, \cdots, n,\ k = 1, \cdots, n$, and take matrices $c^{(k)} = (b_{ki}^* b_{kj}) = (a_i^{(k)*} a_j^{(k)})_{n\times n}$ $(k = 1, 2, \cdots, n)$. Thus, $a = (a_{ij}) = \sum_{k=1}^{n} c^{(k)} = \sum_{k=1}^{n} (a_i^{(k)*} a_j^{(k)})$.

Furthermore, we can get another very useful characterization of positive elements in $M_n(A)$.

Proposition A.10 *Matrix $a = (a_{ij})$ in $M_n(A)$ is positive if and only if $\sum_{i,\ j=1}^{n} x_i^* a_{ij} x_j \ge 0$ for all $x_1 \cdots, x_n$ in A.*

In what follows, we introduce the concept of completely positive mapping:

Definition A.14 Let A and B be C^*-algebras, and let ϕ be a linear mapping from A to B. If the linear mapping $\phi^{(n)}$ from $M_n(A)$ to $M_n(B)$ induced by ϕ such that $\phi^{(n)}(a_{ij}) = (\phi(a_{ij}))$ $((a_{ij}) \in M_n(A))$ is positive, then ϕ is said to be n-positive. Moreover, ϕ is called completely positive if $\phi^{(n)}$ is positive for all positive integer n.

Definition A.15 Let A and B be C^*-algebras, and let ϕ be a linear mapping from A to B. Set $||\phi||_{cb} = \sup\{||\phi^{(n)}|||n = 1, 2, \cdots\}$. If $||\phi||_{cb}$ is finite, then ϕ is called completely bounded. Moreover, if there is still $||\phi|| \le 1$, then ϕ is completely contractive.

It can be proved that if ϕ from A to B is completely positive, then ϕ must be completely bounded and $||\phi||_{cb} = ||\phi||$. Moreover, if ϕ is contractive, then ϕ is completely contractive.

Corollary A.1 *Suppose that A and B are C^*-algebras. If B is also abelian, then every positive linear mapping from A to B is completely positive. In particular, every positive linear functional on A is completely positive.*

It can be further proved the following claim.

Proposition A.11 *Suppose that A and B are C^*-algebras. If A is also abelian, then every positive linear mapping from A to B is completely positive.*

Example A.10 Suppose that A, B, and C are C^*-algebras. Then:

(1) If $\phi : A \to B$ is a $*$-homomorphism, then ϕ is completely contractive and completely positive mapping;
(2) If $\phi : A \to B$ and $\psi : B \to C$ are completely positive mappings, then the composition mapping $\psi \circ \phi : A \to C$ is a completely positive mapping;
(3) If $\phi : A \to B$ is completely positive, then $\Phi : A \to B, \ a \to b^*\phi(a)b$ is completely positive for each fixed $b \in B$.

A.6 Fundamentals of von Neumann Algebras

A.6.1 Basic Concepts of von Neumann Algebra

Let H be a Hilbert space. Recall that the weak operator topology on $B(H)$ is the weak vector topology on $B(H)$ induced by the family J_w of linear functionals $\omega_{x,\,y} : B(H) \to \mathbb{C}$ defined by the equation

$$\omega_{x,\,y}(T) =< Tx, \ y > \quad (x, \ y \in H, \ T \in B(H)).$$

The strong operator topology on $B(H)$ has a base of neighborhoods of origin consisting of the following sets:

$$N(0 : x_1, \ \cdots, \ x_n; \varepsilon) = \{T \in B(H) : \ ||Tx_i|| < \varepsilon\} \ (i = 1, \ \cdots, \ n),$$

where $x_1, \ \cdots, \ x_n$ are in H and ε is positive. Thus, the net $\{T_j\}$ is strong operator convergent to the origin if and only if the net $\{T_j x\}$ of vectors in H converges to 0 for each x in H.

From [93] Theorem 5.1.2, the weak and strong operator closures of a convex subset of $B(H)$ coincide. By [93] Theorem 5.1.3, the unit ball $B(H)_1$ of $B(H)$ is weak operator compact.

Definition A.16 A C^*-algebra A acting on a Hilbert space H is said to be a von Neumann algebra provided it is weak closed in $B(H)$.

Recall that for each von Neumann algebra $A \subset B(H)$, we have an isomorphism between Banach spaces, $A \simeq (C_1(H)/\mathfrak{X})^*$, where $C_1(H)$ is the set of trace class operators on H and $\mathfrak{X}$ is the set of trace class operators T for which $Tr(Tx) = 0$ for all $x \in A$. In particular, A has a w^*-topology which will be referred to as the σ-weak topology. In addition, the space $C_1(H)/\mathfrak{X}$ is called the predual of A and is usually denoted by A_*.

For each subset K of $B(H)$, let K' denote the set of all bounded operators on H commuting with every operator in K. Clearly, K' is a Banach algebra of operators containing the identity operator 1. If K is invariant under the $*$-operation, that is, if $x \in K$ implies $x^* \in K$, then K' is a C^*-algebra acting on H, which is closed with respect to the above weak, strong, and σ-weak topologies, respectively.

The double commutation theorem given by von Neumann is fundamental in the whole theory of operator algebra.

Theorem A.4 *Let A be a unital $*$ algebra on a Hilbert space H. Then A is a von Neumann algebra on H if and only if $A = A''$, where A'' is double commutant of A, that is, $A'' = (A')'$.*

Theorem A.5 *If A is a von Neumann algebra, then $S_1 = \{x \in A : ||x|| \leq 1\}$ is weak operator topology compact.*

Theorem A.6 *(Kaplansky) If A is a von Neumann algebra and B is a dense $*$-subalgebra of A with respect to the weak operator topology, then $\{x \in B : ||x|| \leq 1\}$ is weak operator topology dense in $\{x \in A : ||x|| \leq 1\}$.*

Remark A.3 If $A(\subset B(H))$ is a von Neumann algebra, and x is a self-adjoint element of A with spectral measure E, for each bounded Borel measurable function $f : \sigma(x) \to \mathbb{C}$, define $f(x) = \int f dE$. Then the map $f \to f(x)$ extends the continuous functional calculus for x. What is important to note is that the image of this map is contained in A. It follows that A is the weak closure of set consisting of all finite linear combinations of elements in $P(A)$, where $P(A)$ is denoted the set of all projections in A. Furthermore, the supports and the polar decomposition of each element x of A are still in A.

A.6.2 Types of Factors

A factor is a von Neumann algebra A with trivial center $A \cap A' = \mathbb{C}$.

Definition A.17 Let A be a von Neumann algebra and let $e, f \in A$. We say that e and f are equivalent (which is written as $e \sim f$) if there exists a $v \in A$ such that $v^*v = e$ and $vv^* = f$. We say that e is dominated by f (which is written as $e \prec f$) if there exists a $v \in A$ such that $v^*v = e$ and $vv^* \leq f$.

It is easy to check that $\sim$ ia an equivalence relation and that $\prec$ is transitive.

Proposition A.12 *If $e \prec f$ and $f \prec e$, then $e \sim f$.*

Theorem A.7 *(Comparison Theorem). Let $A(\subset B(H))$ be a von Neumann algebra and let $e,\ f$ be in $P(A)$. Then there exists a projection p in the centra of A such that $ep \prec fp$ and $f(1-p) \prec e(1-p.)$*

Definition A.18 A projection e in a von Neumann algebra A is said to be *finite* if $e \sim f \leq e$ implies $e = f$. Otherwise, it is said to be *infinite*. A projection e is said to be *purely infinite* if there is no nonzero finite projection $f \leq e$ in A. If $c\ e$ is infinite for every central projection c in A with $ce \neq 0$, then e is called *properly infinite*. If eAe is abelian, then e is said to be abelian.

Definition A.19 A von Neumann algebra A is said to be finite, infinite, properly infinite, or purely infinite according to the property of the identity 1, respectively.

Definition A.20 A von Neumann algebra A is said to be of type I if every nonzero central projection in A dominates a nonzero abelian projection in A. If A has no nonzero abelian projection and if every nonzero central projection in A dominates a nonzero finite projection in A, then it is said to be of type II. Furthermore, if A is finite and of type II, then it is said to be of type II_1. If A is of type II and has no nonzero central finite projection, then it is called type II_∞. In addition, if A is purely infinite, that is, there is no nonzero finite projection in A, then it is called type III.

Theorem A.8 *Every von Neumann algebra A is uniquely decomposable into the direct sum of those of type I, type II, and type III.*

Remark A.4 It can be proved that a factor is either of type I, type II_1, type II_∞, or type III.

Theorem A.9 *If A is a factor of type I, then A is $*$-isomorphic to $B(H)$ for some Hilbert space H.*

Definition A.21 A *trace* on a von Neumann algebra A is a function τ on the positive cone A_+ with values in the extended positive reals $[0, +\infty]$ satisfying the following conditions:

(1) $\tau(x + y) = \tau(x) + \tau(y)$ for all $x,\ y \in A^+$;
(2) $\tau(\alpha x) = \alpha\tau(x)$ for all $\alpha \geq 0$ and all $x \in A^+$;
(3) $\tau(x^*x) = \tau(xx^*)$ for all $x \in A$;

with the usual convention $0(+\infty) = 0$. A trace τ is said to be *faithful* if $\tau(x) > 0$ for any $x > 0$, *semifinite* if every nonzero $x \in A_+$ dominates some $y > 0$ with $\tau(y) < +\infty$, *finite* if $\tau(1) < +\infty$, *normal* if $\tau(\sup x_i) = \sup \tau(x_i)$ for every bounded increasing net $\{x_i\}$ in A_+.

Theorem A.10 *Let A be a von Neumann algebra. Then the following statements are equivalent:*

(1) A is finite;
(2) A admits sufficiently many finite normal traces;
(3) There exists a linear map from A to the center $\mathfrak{L}$ with the properties as follows:

(i) $T(x^*x) = T(xx^*) \geq 0$;
(ii) $T(ax) = aT(x)$ *for all* $a \in \mathfrak{L}$ *and* $x \in A$;
(iii) $T(1) = 1$;
(iv) $T(x^*x) > 0$ *for every nonzero* $x \in A$.

If this is the case, then T is unique and σ-weakly continuous.

A.6.3 Characteristics of Type II_1 Factor

We first give an example of type II factor:

Example A.11 Countable discrete groups give rise to von Neumann algebras. In fact one can associate with a discrete group G a von Neumann algebra $L(G)$ in a canonical way: On the Hilbert space $l^2(G)$, the group G has a natural unitary representation $x \to L_x$, the so-called left regular one, which is given by

$$(L_x\xi)(y) := \xi(x^{-1}y) \ (\xi \in l^2(G), \ x, \ y \in G).$$

The group ring $R(G)$ is the linear hull of the set $\{L_x : x \in G\}$ of unitaries. The group von Neumann algebra $L(G)$ associated with G is by definition the closure of $R(G)$ in the weak operator topology. If the group under consideration is ICC (i.e., all its nontrivial conjugacy classes contain infinitely many elements), then the von Neumann algebra $L(G)$ is a type II_1 factor.

Theorem A.11 *Let* $A \subset B(H)$ *be a type* II_1 *factor. Then there is a unique function* $\rho : P(A) \to [0, 1]$ *with the following properties:*

(1) $e \sim f$ *implies* $\rho(e) = \rho(f)$;
(2) $e \perp f$ *implies* $\rho(e + f) = \rho(e) + \rho(f)$;
(3) $\rho(1) = 1$.

The above ρ is called the dimension function of A.

Proposition A.13 *If B is a subfactor of a type* II_1 *factor A, then there exists a unique trace-preserving conditional expectation* $\Phi : A \to B$.

In fact, let p be the orthogonal projection of $L^2(A, \tau)$ onto $L^2(B, \tau|B)$; it can be proved by making $\Phi = p|A$.

At the end of this subsection, we present the fundamental results of A. Connes [41] in which the equivalence between injectivity and hyperfiniteness for the type II_1 factors is proved.

Definition A.22 A von Neumann algebra A is called *hyperfinite* (which is abbreviated as AFD) if there exists an increasing sequence of finite-dimensional subfactors $\{A_n\}_n$ such that $\overline{\cup_n A_n}^w = A$, where $\overline{\cup_n A_n}^w$ is the weak closure of $\cup_n A_n$.

For example, $\otimes_{i=1}^{\infty} M_2(\mathbb{C})_i$ is a hyperfinite factor. Let F_n denote the free group with n generators. From Example A.11, the group algebra $L(F_n)$ is a type II_1 factor, but one can prove that it is non-hyperfinite type II_1 factor.

Definition A.23 A von Neumann algebra acting on a Hilbert space H is called *injective* if it is the range of a conditional expectation Φ defined on $B(H)$.

Theorem A.12 *Let A be a type II_1 factor. Then A is injectivity if and only if it is hyperfiniteness.*

References

1. Accardi, L., Frigerio, A., Lewis, J.T.: Quantum stochastic processes. Publ. RIMS **18**, 97–133 (1982)
2. Aida, S.: Uniformly positivity improving property, Sobolev inequalities and spectral gap. J. Funct. Anal. **158**, 152–185 (1998)
3. Akemann, C., Pedersen, G.K., Tomiyama, J.: Multipliers of C^*-algebras. J. Funct. Anal. **13**, 277–301 (1973)
4. Albeverio, S., Krohn, R.H.: Dirichlet forms and Markov semigroups on $C*$-algebras. Comm. Math. Phys. **56**, 173–187 (1977)
5. Albeverio, S., Ma, Z., Rockner, M.: A Beurling-Deny type structure theorem for Dirichlet forms on general state space. In: Albeverio, S., Fenstad, J.E., Holden, H., Lindstrom, T. (eds.) Ideas and Methods in Mathematical Analysis. Stochastics and Application. Memorial volume for R.Hoegh-Krohn, vol.I. Cambridge University Press, Cambridge (1992)
6. Dajić, A., Koliha, J.J.: Positive solutions to the equations and for Hilbert space operators. J. Math. Anal. Appl. **333**(2), 567–576 (2007)
7. Arendt, W., Batty, C.J.K., Hieber, M., Neubrander, F.: Vector-valued Laplace Transforms and Cauchy Problems. Birkhäüser, Basel, Boston, Berlin (2001)
8. Arveson, W.: Subalgebras of C^*-algebras. Acta Math. **123**, 141–224 (1969)
9. Arveson, W.: Continuous Analogues of Fock Space, No. 409. Memoirs of the American Mathematical Society, Providence, RI (1989)
10. Atiyah, M.: K-Theory. Benjamin, New York (1967)
11. Barnes, B.A., Murphy, G.J., Smyth, M.R., West, T.T.: Riesz and Fredholm in Banach algebras. In: Research Notes in Mathematics, vol. 67. Pitman, London (1982)
12. Ben-Aroya, A., Regev, O., de Wolf, R.: A hypercontractive inequality for matrix-valued functions with applications to quantum computing and LDC. In: IEEE Symposium on Foundations of Computer Science (FOCS), pp. 477–486 (2008)
13. Beurling, A., Deny, J.: Dirichlet spaces. Proc. Natl. Acad. Sci. U.S.A. **45**, 208–215 (1959)
14. Bhat, B.V.R.: An index theory for quantum dynamical semigroups. Trans. AMS **348**(2), 1–23 (1996)
15. Bialynicki-Birula, I., Mycielski, J.: Uncertainty relations for information entropy in wave mechanics. Comm. Math. Phys. **44**, 129–132 (1975)

16. Biane, P.: Free hypercontractivity. Comm. Math. Phys. **184**, 457–474 (1997)
17. Blackadar, B., Handelman, D.: Dimension functions and traces on C^*-algebras. J. Funct. Anal. **45**, 297–340 (1982)
18. Blackadar, B.: K-Theory for Operator Algebras. Springer, Berlin/Heidelberg, New York (1986)
19. Blechner, D.P., Ruan, Z., Sinclair, A.M.: A characterization of operator algebras. J. Funct. Anal. **89**, 188–201 (1990)
20. Blechner, D.P.: A generalization of Hilbert modules. J. Funct. Anal. **136**, 365–421 (1996)
21. Blechner, D.P., Muhly, P.S., Na, Q.: Morita equivalence of operator algebras and their C^*-envelopes. Bull. Lond. Math. Soc. **31**, 581–599 (1999)
22. Blechner, D.P., Muhly, P.S., Paulsen, V.: Categories of operator modules (Morita equivalence and projective modules). In: Memoirs of the AMS, vol.143, Number 681. American Mathematical Society, Providence, Rhode Island (2000)
23. Blumenthal, R.M., Getoor, R.K.: Markov Processes and Potential Theory. Academic Press, New York (1968)
24. Bożejko, M., Speicher, R.: An example of a generalized Brownian motion. Comm. Math. Phys. **137**, 519–531 (1991)
25. Bożejko, M., Speicher, R.: Completely positive maps on Coxeter groups, deformed commutation relations, and operator spaces. Math. Ann. **300**, 97–120 (1994)
26. Bożejko, M., Kümmerer, B., Speicher, R.: q-Gaussian processes: non-commutative and classical aspects. Comm. Math. Phys. **185**, 129–154 (1997)
27. Bożejko, M.: Demonstratio Math. **XLV**, 129–154 (2012)
28. Bratteli, O., Robinson, D.W.: Operator algebras and quantum statistical mechanics I, C^*- and W^*-algebras. In: Symmetry Groups, Composition of States. Springer, Berlin/Heidelberg, New York (1979)
29. Bratteli, O., Robinson, D.W.: Operator algebras and quantum statistical mechanics II. In: Equilibrium States, Models in Quantum Statistical Mechanics. Springer, Berlin/Heidelberg, New York (1981)
30. Breuer, M.: Fredholm theories in von Neumann algebras I. Math. Ann. **178**, 243–254 (1968)
31. Breuer, M.: Fredholm theories in von Neumann algebras II. Math. Ann. **180**, 313–325 (1969)
32. Brown, L.G.: Stable isomorphism of hereditary subalgebras of C^*-algebras. Pac. J. Math. **71**(2), 335–348 (1977)
33. Brown, L., Green, P., Rieffel, M.: Stable isomorphism and strong Morita equivalence of C^*-algebras. Pac. J. Math. **71**, 349–363 (1977)
34. Carlen, E.A., Lieb, E.H.: Optimal hypercontractivity for Fermi fields and related noncommutative integration inequalities. Commun. Math. Phys. **155**, 27–46 (1993)
35. Dong, C.-Z., Wang, Q.-W., Zhang, Y.-P.: The common positive solution to adjointable operator equations. J. Math. Anal. Appl. **396**, 670–679 (2012)
36. Chung K.L.: Lectures from Markov Processes to Brownian Motion. Springer, Berlin/Heidelberg, New York (1982)
37. Cipriani, F.: Dirichlet forms and Markovian semigroups on standard forms of von Neumann algebras. J. Funct. Anal. **147**, 259–300 (1997)
38. Cipriani, F., Sauvageot, J.L.: Derivations as square roots of Dirichlet forms. J. Funct. Anal. **201**, 78–120 (2003)
39. Cipriani, F.: Dirichlet forms as Banach algebras and applications. Pac. J. Math. **223**(2), 229–249 (2006)
40. Combes, F., Zettl, H.: Order structures, traces and weights on Morita equivalent C^*-algebras. Math. Ann. **265**, 67–81 (1983)
41. Connes, A.: Classification of injective factors. Ann. Math. **104**, 73–115 (1976)
42. Connes, A.: Noncommutative Geometry. Academic Press, New York (1994)
43. Cuntz, J.: Simple C^*-algebras generated by isometries. Commun. Math. Phys. **57**, 173–185 (1977)

44. Cuntz, J., Higson, N.: Kuiper's theorem for Hilbert modules. In: Jorgensen, P.E.T., Muhly, P.S. (eds.) Operator Algebras and Mathematical Physics, Contemporary Mathematics, vol. 62, pp. 429–434 (1987). AMS
45. Cvetković-Ilić, D., Wang, Q.-W., Xu, Q.: Douglas' + Sebesty é n's lemmas=a tool for solving an operator equation problem. J. Math. Anal. Appl. **482**, 123599 (2020)
46. Davies, E.B.: Quantum Theory of Open Systems. Academic Press, London, New York, San Francisco (1976)
47. Davies, E.B., Lindsay, J.M.: Non-commutative symmetric Markov semigroups. Math. Z. **210**, 379–411 (1992)
48. Han, D., Larson, D.R.: Bases and group representations. In: Memoirs of the American Mathematical Society, Number 697 (2000)
49. Dixmier, J.: C^*-algebras. North-Holland, Amsterdam (1982)
50. Douglas, R.G.: On majorization, factorization, and range inclusion of operators on Hilbert space. Proc. AMS **17**, 413–415 (1966)
51. Douglas, R.G.: Banach Algebras Techniques in Operator Theory. Academic Press, New York (1972)
52. Douglas, R.G.: On the C^*-algebra of a one-parameter semigroup of isometries. Acta Math. **128**, 143–152 (1972)
53. Echterhoff, S., Quigg, J.: Induced coactions of discrete groups on C^*-algebras. Canad. J. Math. **51**(4), 745–770 (1999)
54. Effros, E.G., Ruan, Z.: Representations of operator bimodules and their applications. J. Oper. Theory **19**, 137–157 (1988)
55. Exel, R.: Morita-Rieffel equivalence and spectral theory for integrable automorphism groups of C^*-algebras. J. Funct. Anal. **172**, 404–465 (2000)
56. Exel, R.: Circle actions on C^*-algebras, Partial automorphisms,and a generalized Pimsner-Voiculescu exact sequence. J. Funct. Anal. **122**, 361–401 (1994)
57. Fagnola, F., Moncino, M.: Free noise dilation of semigroups of countable state Markov processes. Quan. Probab. Rel. Top. **VII**, 149–163 (1992)
58. Fagnola, F.: Unitarity of solutions to quantum stochastic differential equations and conservativity of the associated semigroups. Quan. Probab. Rel. Top. **VII**, 139–148 (1992)
59. Fagnola, F., Rebolledo, R.: Transience and recurrence of quantum Markov semigroups. Probab. Theory Relat. Fields **126**, 289–306 (2003)
60. Farid, F.D., Moslehian, M.S., Wang, Q.-W., Wu, Z.-C.: On the Hermitian solutions to a system of adjointable operator equations. Linear Algebra Appl. **437**, 1854–1891 (2012)
61. Wang, F.: Functional Inequalities, Markov Semigroups and Spectral Theory. Mathematics Monograph Series 4. Science Press, Beijing (2005)
62. Frank, M.: Self-duality and C^*-reflexivity of Hilbert C^*-moduli. Z. Anal. Anwendungen **9**, 165–176 (1990)
63. Frank, M.: Geometrical aspects of Hilbert C^*-modules. Positivity **3**, 215–243 (1999)
64. Frank, M., Larson, D.R.: A module frame concept for HilbertC^*-modules. In: Functional and Harmonic Analysis of Wavelets, Contemporary Mathematics, vol. 247, pp. 207–233 (2000)
65. Frank, M., Larson, D.R.: Frames in Hilbert C^*-modules and C^*-algebras. J. Oper. Theory **48**, 273–314 (2002)
66. Frank, M.: Generalized inverses and polar decomposition of unbounded regular operators on Hilbert C^*-modules. J. Oper. Theory **64**(2), 377–386 (2010)
67. Franz, U., Hong, G., Ulrich, F.M., Zhang, H.: Hypercontractivity of heat semigroups on free quantum groups. J. Oper. Theory. **77**(1), 61–76 (2017)
68. Fuglede, B., Kadison, R.V.: Determinant theory in finite factors. Ann. Math. **55**(3), 520–530 (1952)
69. Fukushima, M.: Dirichlet Forms and Markov Processes. North-Holland, Kodansha (1980)
70. Fukushima, M., Oshima, Y., Takeda, M.: Dirichlet Forms and Symmetric Markov Processes. de Gruyter Studies in Mathematics, Berlin (1994)
71. Goldstein, J.A.: Semigroups of Linear Operators and Applications. Oxford Mathematics Monographs. Oxford University Press, New York (1985)

72. Goldstein, S., Lindsay, J.M.: Beurling Deny conditions for KMS-symmetric dynamical semigroups. C. R. Acad. Sci. Paris. Ser. I **317**, 1053–1057 (1993)
73. Goswami, D., Sinha, K.B.: Hilbert modules and stochastic dilation of a quantum dynamical semigroup on a von Neumann algebra. Commun. Math. Phys. **205**, 377–403 (1999)
74. Gross, L.: Existence and uniqueness of physical ground states. J. Funct. Anal. **10**, 52–109 (1972)
75. Gross, L.: Hypercontractivity and logarithmic Sobolov inequalities for the Clifford Dirichlet forms. Duck Math. J. **42**(3), 383–396 (1975)
76. Gross, L.: Logarithmic Sobolev inequalities. Amer. J. Math. **97**, 1061–1083 (1975)
77. Guido, D., Isola, T., Scarlatti, S.: Non-symmetric Dirichlet Forms on Semifinite von Neumann Algebras. J. Func. Anal. **135**, 50–75 (1996)
78. Hamana, M.: Injective envelopes of C^*-algebras. J. Math. Soc. Jpn. **31**(1), 181–197 (1979)
79. Harte, R.E., Mbekhta, M.: On generalized inverses in C^*-algebras, I. Stud. Math. **103**, 71–77 (1992)
80. Harte, R.E., Mbekhta, M.: Generalized inverses in C^*-algebras, II. Stud. Math. **106**, 129–138 (1993)
81. Heo, J.: Completely multi-positive linear maps and representations on Hilbert C^*-modules. J. Oper. Theory **41**, 3–22 (1999)
82. Heo, J.: Reproducing kernel Hilbert C^*-modules and kernels associated with cocycles. J. Math. Phys. **49**, 1–12 (2008)
83. Hida, T.: Browanian Motion Springer, Berlin/Heidelberg, New York (1980)
84. Lin, H.: The structure of quasi-multipliers of C^*-algebras. Trans. AMS. **315**(1), 147–140 (1989)
85. Lin, H., Hilbert C^*-modules and bounded module mappings. Sci. Sinica (Math.) **12**, 1243–1252 (1990)
86. Lin, H.: Equivalent open projections and corresponding hereditary C^*-subalgebras. J. Lond. Math. Soc. **41**(2), 295–301 (1990)
87. Lin, H., Bounded module maps and pure completely positive maps. J. Oper. Theory **26**, 121–138 (1991)
88. Lin, H.: Injective Hilbert C^*-modules. Pac. J. Math. **154**(1), 131–164 (1992)
89. Lin, H.: Homomorphisms from $C(X)$ into C^*-algebras. Canad. J. Math. **49**, 963–1009 (1997)
90. Lin, H.: An Introduction to the Classification of Amenable C^*-algebras. World Scientific, New Jersey, London, Singapore, Hong Kong, Bangalore (2001)
91. Jensen, K.K., Thomsen, K.: Elements of $K\ \ K$-theory. Birkhauser, Boston, Basel, Berlin (1991)
92. Junge, M., Palazuelos, C., Parcbt, J., Perrin, M., Ricard, E.: Hypercontractivity for free products. Ann. Sci. Ecole. Norm. Sup. **48**(4), 861–889 (2015)
93. Kadison, R.V., Ringrose, J.R.: Fundamentals of the Theory of Operator Algebras I. In: Elementary Theory. Academic Press, New York (1983)
94. Kadison, R.V., Ringrose, J.R.: Fundamentals of the theory of operator algebras II. In: Advanced Theory. Academic Press, New York (1986)
95. Kajiwara, T., Pinzari, C., Watatani, Y.: Ideal structure and simplicity of the C^*-algebras generated by Hilbert bimodules. J. Funct. Anal. **159**, 295–322 (1998)
96. Kaplan, A.: Covariant completely positive maps and liftings, Rocky Mountain. J. Math. **23**(3), 939–945 (1993)
97. Kaplansky, I.: Modules over operator algebras. Amer. J. Math. **75**, 839–853 (1953)
98. Kasparov, G.G.: Hilbert C^*-modules: theorems of Stinespring and Voiculescu. J. Oper. Theory **4**, 133–150 (1980)
99. Kasparov, G.G.: The operator K-functor and extensions of C^*-algebras. Izv. Akad. Nauk SSSR. Ser. Math. **44**(3), 571–636 (1980)
100. Katsoulis, E.G., Kribs, D.W.: Tensor algebras of C^*-correspondences and their C^*-envelopes. J. Funct. Anal. **234**, 226–233 (2006)
101. Kato, T.: Perturbation Theory for Linear Operators. Springer, Berlin/Heidelberg, New York (1966)

102. Kosaki, H.: Applications of uniform convexity of noncommutative L^p spaces. Trans. AMS. **283**(1), 265–282 (1984)
103. Królak, I.: Optimal holomorphic hypercontractivity for CAR algebras. Bull. Pol. Acad. Sci. **58**, 79–90 (2010)
104. Kuiper, N.: The homotopy type of the unitary group of Hilbert space. Topology **3**, 19–30 (1965)
105. Kummerer, B.: Markov dilations on W^*-algebras. J. Funct. Anal. **63**, 139–177 (1985)
106. Kuo, H.: Gaussian measures in Banach spaces. In: Lecture Notes in Mathematics, vol. 403. Springer, Berlin/Heidelberg, New York (1975)
107. Kusuda, M.: Characterizations of certain classes of hereditary C^*-subalgebras. Proc. AMS. **116**(4), 999–1005 (1992)
108. Laca, M.: Endomorphisms of $B(H)$ and Cuntz algebras. J. Oper. Theory **30**, 85–108 (1993)
109. Lance, E.C.: Hilbert C*-modules:A Toolkit for Operator Algebraists. London Mathematical Society, Lecture Note Series 124. Cambridge University Press, Cambridge (1994)
110. Lance, E.C.: Unitary operator on Hilbert C^*-modules. Bull. Lond. Math. Soc. **26**, 363–366 (1994)
111. Lant, T., Thieme, H.R.: Markov transition functions and semigroups of measures. Semigroup Forum **74**, 337–369 (2007)
112. Lesch, M.: On the index of the infinitesimal generator of a flow. J. Oper. Theory **25**,73–92 (1991)
113. Bengren, L.: Operators Algebras. In: Basic Modern Mathematics Series. China Science Press, Beijing (1986)
114. Bengren, L.: Banach Algebras. China Science Press, Beijing (1992)
115. Lindsay, J.M.: Gaussian hypercontractivity revisited. J. Funct. Anal. **92**, 313–324 (1990)
116. Lindsay, J.M., Meyer, P.A.: Fermionic hypercontractivity. In: Quantum Probability VII. World Scientific, Singapore (1992)
117. Loring, T.A.: Projective C^*-algebras. Math. Scand. **731**, 274–280 (1993)
118. Lust-Piquard, F.: Riesz transforms on deformed Fock spaces. Commun. Math. Phys. **205**, 519–549 (1999)
119. Magajna, B.: Hilbert C^*-modules in which all closed submodules are complemented. Proc. AMS. **125**(3), 849–852 (1997)
120. Magajna, B.: A topology for operator modules over W^*-algebras. J. Funct. Anal. **154**, 17–41 (1998)
121. Meyer, P.A.: Quantum Probability for Probabilistits. In: Lectures Notes in Mathematics, vol. 1538. Springer, Berlin/Heidelberg, New York (1993)
122. Gao, M.: Free Markov pocesses and stochastic differential equations in von Neumann algebras. Ill. J. Math. **52**, 153–180 (2008)
123. Mingo, J.A., Phillips, W.J.: Equivariant triviality theorems for Hilbert C^*-modules. Proc. Amer. Math. Soc. **91**, 225–230 (1984)
124. Mingo, J.A.: K-theory and multipliers of stable C^*-algebras. Trans. Amer. Math. Soc. **299**, 397–411 (1987)
125. Mingo, J.A., Spielberg, J.S.: The index of normal Fredholm elements of C^*-algebras. Proc. AMS. **113**(1), 187–192 (1991)
126. Mousavi, Z., Eskandari, R., Moslehian, M.S., Mirzapour, F.: Operator equations and in Hilbert-modules. Linear Algebra Appl. **517**, 85–98 (2017)
127. Muhly, P.S., Solel, B.: Hilbert modules over operator algebras. Mem. Amer. Math. Soc. 117(559). Providence, Rhode Island (1995)
128. Muhly, P.S., Solel, B.: Tensor algebras over C^*-correspondences: representations, Dilations, and C^*-envelopes. J. Funct. Anal. **158**, 389–457 (1998)
129. Muhly, P.S., Solel, B.: On the Morita equivalence of tensor alhgebras. Proc. Lond. Math. Soc. **81**(3), 113–168 (2000)
130. Murphy, G.J.: C^*-algebras and Operator Theory. Academic Press, London (1990)
131. Murphy, G.J.: Positive definite kernels and Hilbert C^*-modules. Proc. Edinb. Math. Soc. **40**, 367–374 (1997)

132. Murphy, G.J.: Fredholm index theory and the trace. Proc. R. Ir. Acad. **94A**(2), 161–166 (1994)
133. Nelson, E.: A quartic interaction in two dimensions. In: Mathematical Theory of Elementary Particles. pp. 67–73. MIT Press, Cambridge (1965)
134. Nelson, E.: Notes on non-commutative Integration. J. Funct. Anal. **15**, 103–116 (1974)
135. Olkiewicz, R., Zegarlinski, B.: Hypercontractivity in noncommutative L^p spaces. J. Funct. Anal. **161**(1), 246–285 (1999)
136. Park, Y.M.: Construction of Dirichlet Forms on Standard Forms of von Neuman algebras. Infin. Dimens. Anal. Quantum Probab. Relat. Top. **3**(1), 1–14 (2000)
137. Paschke, W.L.: Inner product modules overB^*-algebras. Trans. Amer. Math. Soc. **182**, 443–468 (1973)
138. Paulsen, V.I.: Completely bounded maps and dilations. Pitman Research Notes in Mathematical Series, vol. 146. Longman, Harlow, Essex, Britain (1986)
139. Pazy, A.: Semigroups of Linear Operators and Applications to Partial Diffential Equations. Springer, Berlin/Heidelberg, New York (1983)
140. Pedersen, G.K.: C^*-algebras and Their Automorphism Groups. Academic Press, London (1979)
141. Pimsner, M., Popa, S., Voiculescu, D.: Honogeneous extensions of $C(X) \otimes K(H)$ II. J. Oper. Theory **4**, 211–249 (1980)
142. Pimsner, M.V.: A class of C^*-algebras generalizing both Cuntz-Krieger algebras and crossed products by $\mathbb{Z}$. Fields Institute Commun. **12**, 189–212 (1997)
143. Popovici, D.: Minimal unitary dilations of contractions on Hilbert C^*-modules, S.L.O.H.A., West University of Timispoara, (1996)
144. Popovici, D.: Orthogonal decompositions of isometries in Hilbert C^*-modules. J. Oper. Theory **39**, 99–112 (1998)
145. Wang, Q.-W., Dong, C.-Z.: The positive solution to a system of adjointable operator equations over Hilbert C^*-modules. Linear Algebra Appl. **433**, 1481–1489 (2010)
146. Xu, Q.: Common Hermitian and positive solutions to the adjointable operator equations $AX = C, XB = D$. Linear Algebra Appl. **429**, 1–11 (2008)
147. Xu, Q., Sheng, L.: Positive semi-definite matrices of adjointable operators on Hilbert C^*-modules. Linear Algebra Appl. **428**(4), 992–1000 (2008)
148. Xu, Q., Sheng, L., Gu, Y.: The solutions to some operator equations. Linear Algebra Appl. **429**, 1997–2024 (2008)
149. Accardi, L., Fagnola, F. (eds.): Quantum interacting particle systems. In: QP-PQ, Quantum Probability and White Noise Analysis, vol. XIV. World scientific, New Jersey, London, Singapore, Hong Kong (2002)
150. Reed, M., Simon, B.: Methods of modern mathematical physics I. In: Functional Analysis. Academic Press, New York (1972)
151. Reed, M., Simon, B.: Methods of modern mathematical physics II. In: Fourier Analysis. Academic Press, New York (1975)
152. Ricard, E., Xu, Q: A noncommutative martingale convexity inequality. Ann. Probab. **44**(2), 867–882 (2016)
153. Rieffel, M.A.: Morita equivalence for C^*-algebras and W^*-algebras. J. Pure Appl. Algebras **5**, 51–96 (1974)
154. Rieffel, M.A.: Induced representations of C^*-algebras. Adv. Math. **13**, 176–257 (1974)
155. Rieffel, M.A.: Morita equivalence for operator algebras. In: Kadison, R.V. (ed.) Operator Algebras and Applications, Proceedings of Symposium Pure Mathematics, vol. 38, Part I, pp. 285–298. American Mathematical Society (1982)
156. Roch, S., Silbermann, B.: C^*-algebra techniques in numerical analysis. J. Oper. Theory **35**, 241–280 (1996)
157. Röckner, M., Wang, F.-Y.: Weak Poincaré inequalities and L^2-convergence rates of Markov semigroups. J. Funct. Anal. **185**, 564–603 (2001)
158. Sauvageot, J.L.: Markov quantum semigroups admit covariant Markov C^*-dilations. Commun. Math. Phys. **106**, 91–103 (1986)

159. Sauvageot, J.L.: Quantum Dirichlet forms, differential calculus and semigroups. In: Accardi, L., von Waldenfels, W. (eds.) Quantum Probability and Applications V. Lecture Notes in Mathematics, vol. 1442. Springer, Berlin/Heidelberg, New York (1988)
160. Sauvageot, J.L.: Strong Feller semigroups on C^*-algebras. J. Oper. Theory **42**, 83–102 (1999)
161. Schweizer, J.: Hilbert C^*-modules with a predual. J. Oper. Theory **48**, 621–632 (2002)
162. Segal, I.E.: A non-commutative extension of abstract integration. Ann. Math. **57**, 401–456 (1953)
163. Zhang, S.: Certain C^*-algebras with real rank zero and the internal structure of their corona and multiplier algebras, Part I. Pac. J. Math. **115**(1), 169–179 (1992)
164. Zhang, S.: A Riesz decomposition property and ideal structure of multiplier algebras. J. Oper. Theory **24**, 209–225 (1990)
165. Zhang, S.: C^*-algebras with real rank zero and the internal structure of their corona and multiplier algebras III. Canad. J. Math. **42**, 159–190 (1990)
166. Zhang, S.: Exponential rank and exponential length of operators on Hilbert C^*-modules. Ann. Math. **137**, 129–144 (1993)
167. Silverstein, M.L.: Symmetric Markov processes. In: Lecture Notes in Mathematics, vol. 426. Springer, Berlin/Heidelberg, New York (1974)
168. Skeide, M.: Hilbert modules in quantum electro dynamics and quantum probability. Commun. Math. Phys. **192**, 569–604 (1998)
169. Skeide, M.: Generalised matrix C^*-algebras and representations of Hilbert modules. Math. Proc. R. Ir. Acad. **100**, 11–38 (2000)
170. Skeide, M.: Quantum stochastic calculus on full Fock modules. J. Funct. Anal. **173**, 401–452 (2000)
171. Skeide, M.: The Index of White Noises and Their Product Systems. Centro Vito Volterra, Rome (2001)
172. Skeide, M.: Dilations, product systems, and weak dilations. Math. Notes **71**(6), 836–843 (2002)
173. Smith, R.R.: Completely bounded maps between C^*-algebras. J. Lond. Math. Soc. **27**(2), 157–166 (1983)
174. Solel, B.: Operator algebras over C^*-correspondences. In: Operator Algebras and Applications, Samos, 1996. NATO Advanced Science Institutes Series C: Mathematical and Physical Sciences, vol. 495, pp.429–448. Kluwer Academic, Dordrecht, Norwell, MA. (1997)
175. Speicher, R.: Combinatorial Theory of the Free Product with Amalgamation and Operator-Valued Free Probability Theory, vol. 627. Memoirs of the AMS (1998)
176. Stinespring, F.: Positive functions on C^*-algebras. Proc. AMS **6**, 211–216 (1955)
177. Stinespring, W.F.: Integration theorems for gages and duality for unimodular groups. Trans. Amer. Soc. **90**, 15–56 (1959)
178. Sz-Nagy, B., Foias, C.: Dilatations des commutants d'operateurs. C. R. Acad. Sci. Paris Ser. A **266**, 493–495 (1968)
179. Sz-Nagy, B., Foias, C.: Harmonic Analysis of Operators on Hilbert Space. North-Holland, Amsterdam (1970)
180. Takesaki, M.: Theory of Operator Algebras I. Springer, Berlin/Heidelberg, New York (1979)
181. Toda, M., Kubo, R., Saitô, N.: Statistical Physics, I, II. Springer, Berlin/Heidelberg, New York (1992)
182. Tsui, S.: Completely positive module maps and completely positive extreme maps. Proc. AMS **121**(2), 437–445 (1996)
183. Watatani, Y.: Index for C*-subalgebras, vol. 83. Memoirs of Americal Mathmatical Society (1990)
184. Wegge-Olsen, N.E.: K-theory and C^*-algebras. Oxford University Press (1993)
185. Liang, W., Deng, C.: The solutions to some operator equations with corresponding operators not necessarily having closed ranges. Linear Multilinear Algebra **67**(8), 1606–1624 (2019)
186. Woronowicz, S.L.: Unbounded elementes affiliated withC^*-algebras and non-compact quantum groups. Commun. Math. Phys. **136**, 399–432 (1991)
187. Woronowicz, S.L.: Compact Quantum Groups. Warsaw, North-Holland, Amsterdam (1998)

188. Fang, X., Yu, J.: Solutions to operator equations on Hilbert $C*$-modules II. Integral Equs. Oper. Theory **68**, 23–60 (2010)
189. Yeadon, F.J.: Non-commutative L^p-spaces. Math. Proc. Camb. Phil. Soc. **56**, 173–187 (1977)
190. Yosida, K.: Functional Analysis. Springer, Berlin/Heidelberg, New York (1968)
191. Gongqing, Z., Maozheng, G.: Lecture Notes on Functional Analysis. Peking University Press, Beijing (1990)
192. Lunchuan, Z.: The characterization of bounded generalized inverse module maps and applying in C^*-algebras. Abstracts of short communications and poster sessions. In: Operator Algebras and Functional Analysis, P_{164}, ICM2002. Higher Education Press, Beijing (2002)
193. Lunchuan, Z.: Complemented Hilbert C^*-modules and bounded module mappings. Adv. Math. (Chin.) **31**(3), 275–278 (2002)
194. Lunchuan, Z.: Relation between Hereditary C^*-subalgebras and complemented submodules. Acta Math. Sinica (Chin.) **47**(4), 747–750 (2004)
195. Lunchuan, Z.: Bounded generalized inverse module mappings and applications. Acta Math. Sinica (Chin.) **49**(1), 7–10 (2006)
196. Lunchuan, Z.: The factor decomposition theorem of bounded generalized inverse module maps. Acta Math. Sinica, English Ser. **23**(8), 1413–1418 (2007)
197. Lunchuan, Z., Maozheng, G.: Stone type theorem. Acta Math. Sinica (Chin.) **50**(4), 857–860 (2007)
198. Lunchuan, Z.: The characterization of Moore-Penrose inverse module maps. Rocky Mountain J. Math. **38**(1), 351–357 (2008)
199. Lunchuan, Z., Maozheng, G.: The characterization of certain quantum stochastic stationary process. Acta Math. Sinica, English Ser. **26**(9), 1807–1814 (2010)
200. Lunchuan, Z., Maozheng, G.: The characterization of a class of quantum Markov semigroups and associated operator-valued Dirichlet forms based on Hilbert C^*-modules, Abstracts of short communications and poster sessions. In: Operator Algebras and Functional Analysis, ICM 2010, India (2010)
201. Lunchuan, Z., Maozheng, G.: Unitary equivalence between closed submodules and the problem of hereditary C^*-subalgebras. Acta Math. Sinica (Chin.) **53**(6), 1041–1044 (2010)
202. Lunchuan, Z., Maozheng, G.: The characterization of a class of quantum Markov semigroups and the associated operator-valued Dirichlet forms based on Hilbert C*-module. Sci. China Math. **57**(2), 377–387 (2014)
203. Lunchuan, Z., Maozheng, G.: Beurling-Deny correspondence of a class of quantum Markov semigroups and the associated operator-valued Dirichlet forms. In: ICM 2014. Seoul (2014)
204. Lunchuan, Z.: Equivalence between log-Sobolev inequality and hypercontractivity of quantum Markov semigroup. In: ICM 2018. Rio de Janeiro, Brazil (2018)
205. Lunchuan, Z.: Characterization of a class of Markov module operator semigroups and the corresponding operator-valued Dirichlet forms. Chin. J. Comtemp. Math. **39**(2), 171–180 (2018)
206. Lunchuan, Z.: The Equivalence of Hypercontractivity and Logarithmic Sobolev Inequality for q $(-1 < q < 1)$-Ornstein-Uhlenbeck Semigroup. Chin. Ann. Math. **414**(4), 615–626 (2020)
207. Lunchuan, Z.: Operator-valued Dirichlet forms and module operator Markov semigroups, Chapter 11, 161–179. In: Zheng, Z. (ed.) Proceedings of the First International Forum on Financial Mathematics and Finacial Technology. Springer, Berlin/Heidelberg, New York (2020)
208. Lunchuan, Z.: Equivalence between logarithmic Sobolev inequality and hypercontractivity in a probability gage space. Proc. Edinb. Math. Soc. **64**(1), 59–71 (2021)
209. Ma, Z., Rockner, M.: An Introduction to the Theory of (Non-symmetric) Dirichlet Forms, Universitext. Springer, Berlin/Heidelberg, New York (1992)
210. Ma, Z., Rockner, M.: Markov processes associated with positivity preserving coercive forms. Canad. J. Math. **47**(4), 817–840 (1995)
211. Ruan, Z.: Subspaces of C^*-algebras. J. Funct. Anal. **76**, 217–230 (1988)
212. Ruan, Z.: Injectivity of operator spaces. Trans. AMS. **315**(1), 89–104 (1989)

Index

L. Zhang, *Hilbert C*- Modules and Quantum Markov Semigroups*,
https://doi.org/10.1007/978-981-99-8668-2